Lecture Notes in Business Information Processing

293

More information about this series at http://www.springer.com/series/7911

Mareike Schoop · D. Marc Kilgour (Eds.)

Group Decision and Negotiation

A Socio-Technical Perspective

17th International Conference, GDN 2017
Stuttgart, Germany, August 14–18, 2017
Proceedings

 Springer

Editors
Mareike Schoop
University of Hohenheim
Stuttgart
Germany

D. Marc Kilgour
Wilfrid Laurier University
Waterloo, ON
Canada

ISSN 1865-1348   ISSN 1865-1356   (electronic)
Lecture Notes in Business Information Processing
ISBN 978-3-319-63545-3   ISBN 978-3-319-63546-0   (eBook)
DOI 10.1007/978-3-319-63546-0

Library of Congress Control Number: 2017948292

Printed on acid-free paper

This Springer imprint is published by Springer Nature
The registered company is Springer International Publishing AG
The registered company address is: Gewerbestrasse 11, 6330 Cham, Switzerland

Preface

The field of Group Decision and Negotiation focuses on decision processes with at least two participants and a common goal but conflicting individual goals. Group decisions and negotiations can be performed in both an intra-organisational as well as an inter-organisational context. They consist of complex processes, including preference elicitation, preference adjustment, proposals and counter-proposals, and choice. Communication and decision-making are key to group decision and negotiation processes; sophisticated support for these functions is thus a central objective of the Group Decision and Negotiation field.

Research areas of Group Decision and Negotiation include electronic negotiations, experiments, the role of emotions in group decision and negotiation, preference elicitation and decision support for group decisions and negotiations, and conflict resolution principles.

The 17th International Conference on Group Decision and Negotiation (GDN 2017) continues the long history of GDN conferences as the primary forum for researchers and practitioners in the fields of group decision and negotiation. GDN 2017 is a truly international conference, with participants from Europe, the Americas, Asia, Africa, and Oceania. Especially in times of global conflict and uncertainty, state-of-the-art research on dealing with conflicts in a cooperative and integrative way is more important than ever.

The book contains 14 full papers chosen from 85 submissions to GDN 2017. The first two papers are authored by the keynote speakers Matthias Jarke and Dov Te'eni. The remaining papers are organised in several streams that demonstrate the variety of research successes presented at GDN 2017:

- The stream on "General Topics in Group Decision and Negotiation" includes papers covering a broad range of topics from across the GDN field, from formal foundations to practical applications.
- The "Conflict Resolution" stream analyses strategic conflicts between individuals and groups in diverse application areas. The role of information technology in general, and dedicated systems in particular, are assessed.
- The "Emotions in Group Decision and Negotiation" stream examines the subjective and inter-subjective role of emotions affecting group decisions and negotiations.
- The "Negotiation Support Systems and Studies" stream focuses on electronic negotiations using several systems and tools, and includes system designs and laboratory and field studies analysing e-negotiations, mediation, and facilitation.
- The "Preference Modelling for Group Decision and Negotiation" stream focuses on approaches supporting groups of negotiators and decision-makers in eliciting goals and preferences and on scoring systems for assessing offers.

Organising an international conference on Group Decision and Negotiation certainly requires many negotiations among many parties, and a great deal of cooperative group decision-making.

We are very pleased at how well the conference has come together. We would particularly like to thank:

- The three keynote speakers Wendi Adair, Matthias Jarke, and Dov Te'eni, for providing stimulating, innovative, and challenging research insights;
- The Conference Chairs, for regular interactions and advice;
- The Organising Chairs and the Organising Committee, for their work in putting together this splendid conference;
- The authors of the 94 papers submitted to the conference and the doctoral consortium;
- The members of the Programme Committee and the reviewers, for providing careful feedback and comments on all papers;
- Springer, for providing the funds for the Best Paper Awards;
- The University of Hohenheim and the Faculty of Business, Economics, and Social Sciences, and especially their research area on Negotiation Research (NegoTrans), for their support and for making it possible for the conference to take place in the beautiful venue of Hohenheim Castle;
- Hohenheim Management Development e.V., and Unibund, for their generous financial support.

August 2017   Mareike Schoop
                                                                 D. Marc Kilgour

Organisation

Programme Chairs

Mareike Schoop   University of Hohenheim, Germany
Marc Kilgour   Wilfried Laurier University, Canada

Conference Chairs

Rudolf Vetschera   University of Vienna, Austria
Pascale Zaraté   University of Toulouse 1 Capitole, France

Honorary Conference Chair

Melvin F. Shakun   New York University, USA

Organising Chairs

Annika Lenz   University of Hohenheim, Germany
Philipp Melzer   University of Hohenheim, Germany

Doctoral Consortium Chairs

Per van der Wijst   Tilburg University, The Netherlands
Tomasz Wachowicz   University of Economics in Katowice, Poland

Programme Committee

Fran Ackerman   University of Strathclyde, UK
Adiel Almeida   Federal University of Pernambuco, Brazil
Reyhan Aydogan   Delft University of Technology, The Netherlands,
       and Ozyegin University, Turkey
Deepinder Bajwa   Western Washington University, USA
Martin Bichler   Technical University of Munich, Germany
Tung Bui   University of Hawai'i, USA
João C. Clímaco   Coimbra University, Portugal
Xusen Cheng   University of International Business and Economics,
       China
Suzana F.D. Daher   Federal University of Pernambuco, Brazil
Gert-Jan de Vreede   University of South Florida, USA
Luis Dias   University of Coimbra, Portugal
Colin Eden   University of Strathclyde, UK
Love Ekenberg   Stockholm University, Sweden

Organising Committee

Corina Blum	University of Hohenheim, Germany
Muhammed-Fatih Kaya	University of Hohenheim, Germany
Michael Körner	University of Hohenheim, Germany
Andreas Schmid	University of Hohenheim, Germany
Bernd Schneider	University of Hohenheim, Germany
Serkan Sepin	University of Hohenheim, Germany

Contents

Preference Modelling for Group Decision and Negotiation

Keynote Papers

University of Hohenheim

Data Spaces: Combining Goal-Driven and Data-Driven Approaches in Community Decision and Negotiation Support

Matthias Jarke[1,2(✉)] [iD]

[1] Information Systems, RWTH Aachen University,
Ahornstr. 55, 52074 Aachen, Germany
jarke@dbis.rwth-aachen.de
[2] Fraunhofer FIT, Schloss Birlinghoven, 53754 Sankt Augustin, Germany

Abstract. In the last decade, social network analytics and related data analysis methodologies have helped big players gain enormous influence on the web, largely due to clever centralistic data collection in major data lakes. In the form of recommender systems, this can also be seen as world-scale group decision support. In our research, we have been more interested in how these kinds of technologies can spill over to smaller-scale communities of interest in the long tail of the internet. Examples include learning communities and open source software development communities of individuals, but also questions of controlled data and knowledge sharing among small and medium enterprises or medical institutions. Especially in the latter cases, we often face strongly conflicting goals that need to be negotiated to mutually acceptable solutions, quite along the original GDSS and NSS visions of Mel Shakun and colleagues. One example is medical research support on rare diseases which raises the need for data sharing across multiple health organizations (not necessarily being fond of each other) in a fully transparent, fraud-resistant research process while preserving best-possible privacy of patient data. We end with a summary of the Industrial Data Space initiative recently proposed by Fraunhofer which aims at architectures, rules and tools for data sovereignty in cross-organizational data management and analytics.

Keywords: Data exchange · Requirements engineering · Industrial Data Space · Community decision support

1 Introduction

In 1983, I was fortunate to advise some of the internationally first doctoral theses addressing distributed group decision (Tung Bui [BJ86]) and negotiation support systems (Tawfik Jelassi [JJS87]). Both works were heavily influenced by cooperation with Eric Jacquet-Lagrèze on multi-person multi-criteria decision models and Melvin Shakun's Evolutionary Systems Design approach that interpreted negotiation as a re-definition of the search space for creative solutions. However, case studies conducted at Renault showed the importance of a third factor in the negotiation context: the targeted sharing of data and knowledge among negotiation partners. Almost exactly

© Springer International Publishing AG 2017
M. Schoop and D.M. Kilgour (Eds.): GDN 2017, LNBIP 293, pp. 3–14, 2017.
DOI: 10.1007/978-3-319-63546-0_1

30 years before the present conference, the EJORS journal published our paper on MEDIATOR, a prototype NSS that tried to integrate all three aspects of MCDM, ESD, and data management [JJS87].

Almost at the same time, Terry Winograd and Fernando Flores published their influential book [WF86] in which they called attention to the importance of communication processes and media in the negotiation and execution of joint decisions, based on language action theory. While their originally intended application of these models as the foundation of workflow systems had some success in service application domains such as hospital coordination, supply chain or complaint management [Scha96, QS01], the probably biggest impact was the advent of *service-oriented architectures* at the software level which applied a basic version of their protocol for the coordination of software components – a key ingredient of today's universal digitization.

A decade later, Terry Winograd and his students Sergey Brin and Larry Page initiated the next major step by proposing to exploit these data traces for large-scale social network analytics. Their PageRank algorithm [PBMW99] – originally intended for evaluating the importance of scientists (CiteSeer ranking) – quickly emerged as one of the decisive success factors for their search engine start-up, Google [BP98], together with smart indexing and parallel computing methods. Amazon's recommender systems and Facebook's "friendship" networks are just two of numerous further examples of data-driven group decision and negotiation systems that dominate the debate (and stock market values) due to the economic exploitation of the underlying network effects. Group decision and negotiation support at the end-user level have thus become highly automated data-driven mass phenomena, with data concentrating in the hands of a few very large platform owners.

In the B2B sector, but also in special interest communities of the long tail of the Internet, this development has been causing significant concern, as the data-producing networks will grow again by orders of magnitude through the Internet of Things (sensors, actuators, multimedia cameras, smart watches, etc.). The fear is that the direct contact to customers gets lost, e.g., the customer sees a car as an expensive strange interface to her search engine, entertainment system, or social network. Equally important, the detailed data traces left by users and IoT might yield more confidential product and process knowledge to the exchange platform than to the producer itself.

These observations have led to the creation of multiple special purpose exchange platforms with domain-specific requirements, technological solutions, and business models. Much of our recent research has focused on supporting communities of practice in the long tail of the Internet, such as Open Source development communities [NHKJ16], or learning resp. vocational training communities [PKK15].

In the present paper, we focus on B2B data and knowledge exchange underlying intra-organizational and inter-organizational decisions and related negotiations. In Sect. 2, we report on experiences with three major application domains in order to highlight the requirements and some partial solutions. *Data sovereignty* appears as a joint requirement in all areas. In Sect. 3, we investigate the concept of "data spaces" as a possible solution and contrast it with the currently fashionable "data lake" idea. The paper then ends with an overview of the Fraunhofer Industrial Data Space initiative which tries to address these issues at a large scale from many different perspectives.

2   Three Case Studies

In this section, we illustrate the issues encountered in information and service exchange platforms with group decision or negotiation support aspects with projects conducted in three different application domains: intermodal and cross-organizational personal mobility, telemedicine, and cooperative robotics. The main goal is to illustrate challenges and partial solution options such as many aspects of heterogeneity, value appropriation issues, and the importance of data ownership.

2.1   Mobility Broker Platform

Navigation systems for cars and bikes have become a favorite support tool for individual mobility. Following the arrival of the market-leading TomTom system (which uses massively parallel data analytics on exclusive real-time telephone data to provide better prediction of traffic conditions as a key competitive advantage), public transport systems feel the urgent need to offer similar services, not just for adaptive navigation but also for simpler transaction handling and accounting.

In cooperation with RWTH Aachen University, Deutsche Bahn developed such an infrastructure for their train systems, right in time for the arrival of the smartphone. Following the "mobile first" strategy, the project built the App *DBNavigator*, by now the most downloaded "serious" App in Europe with over 36 million active installations.

However, coverage of the own network is only of limited help for travelers who use an intermodal mix of trams and busses, taxis and other driving services, car sharing and rental eBikes, to optimize their mobility in terms of criteria such as cost, time, or comfort. The advent of eBikes and eCars, even autonomous driving options, adds to this richness of choices but also to further constraints. These different modalities are usually offered by different vendors, and even in the same modality, we may find competing companies in the same or adjacent regions where you want to travel.

Based on the mentioned previous research with the train system a series of projects, in part initiated and coordinated by our group, aim to enable small and medium vendors to join an open one-stop service network. The most recent of these projects, called Mobility Broker [BG*16], resulted firstly in a meta model for semantic data exchange of all relevant data agreed by all regional traffic organizations in the country and with some neighbors. On this homogenized database, advanced services include adaptive, intermodal routine, taking real-time problem information or personal plan changes into account, and a universal ticketing system with significantly lower production cost than previous ones. Advanced payment and accounting options are included as well (cf. Fig. 1). An installation showcasing especially the optimal usage of eMobility has been commercially operational in the city of Aachen since late 2016. Based on this success, the German Federal Ministry of Traffic Infrastructure (BMVI) initiated a program of follow-up projects for a nation-wide extension. Beside the complex agreement on a standardized data exchange model, the definition of suitable business models for service exchange and value appropriation among participating players in such a peer-to-peer setting prove a key challenge we are pursuing based on service-dominant logic [PJ16].

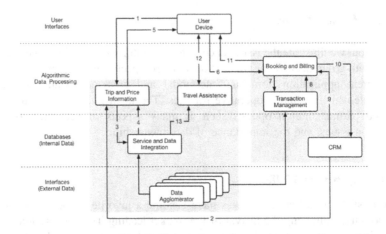

Fig. 1. Service flow and sample components of eMobility Broker components in Aachen

2.2 Information Sharing in Cooperative Telecare

Together with the continuing population moves to large cities, the growing share of elderly people in European populations causes a shortage in medical personnel especially in rural areas. Moreover, most elderly people prefer to stay at their accustomed home rather than moving to special elderly-care homes.

In reaction to such demands, tele-consulting and telecare at home from remote nurses, doctors, or pharmacists are capturing increasing attention in research and practice. Together with the German Society for Telemedicine and industry partners such as Bayer Healthcare, we have developed such a system that aims at creating a social network not just between patient and medical specialists as mentioned above, but also with family, friends, and other patients. From the user interaction perspective, the main design criterion was minimal learning effort and a minimum (preferably none) of new devices; all functions can be easily operated through a standard TV set, but it is also easily possible to link sensor tracking to the system. Besides several public demonstrations, the system has been in successful trials with several hundred patients, and is being rolled out to a five-digit number of diabetes and post-operative patients in Germany (Fig. 2).

Fig. 2. Fraunhofer FIT's telecare environment is demonstrated at the Max-Planck-Fraunhofer Event "Pearls of Research" in the German Chancellery

In contrast, the data management and exchange infrastructure behind such a platform is quite complex. On the one hand, appropriate case information must be made available under strict authentication, security, and privacy precautions, following all the rather demanding corresponding rules. Moreover, these static data may have to be linked to dynamically captured data streams of sensor data captured in real-time or buffered somewhere using communication channels that are not always of good quality. On the user side, competency profiles may be needed for personalization of the interaction style to possible media-related, bodily or mental limitations.

Another goal of the system is to bring together self-assistance communities of people with similar disease patterns, but also to enable research across multiple cases, thus augmenting clinical studies beyond the time of hospitalization. For these studies – given the slate of recent scandals –, the exact opposite of privacy is required, i.e. full transparency of the treatment and research processes. And in yet another angle, all institutional players in this domain have strong interests of their own which often let them hesitate to share data. The complexity of the situation can be seen from Fig. 3, a generalization of the situation found in the telecare case.

2.3 eRobotics

In many engineering applications, "digital twins" are becoming a popular tool for real-time data-driven decision support. Digital twins are decision or simulation models that ideally run in parallel to the real system, such that they can collect tracking data from it to analyze its status, simulate possible alternatives for the next steps, and play them back into the control of the real system. Obviously, to achieve this, the digital twin has to operate faster than the real system – an extreme demand if the system to be

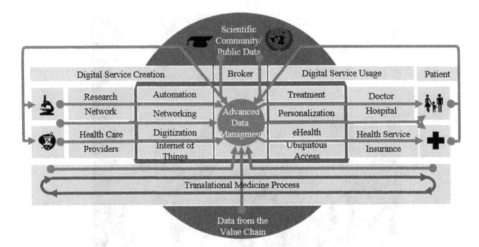

Fig. 3. The Medical Data Space aims at linking operational medical and administrative processes to medial research in a transparent, yet privacy-conserving translational medicine [FJ16]

controlled is distributed, managed by multiple specialists from multiple viewpoints in heterogeneous viewpoints, and requires 4D simulation in space and time to be realistic.

We find such a setting especially in robotics applications in hostile environments, e.g. when an unmanned cargo space ship wants to dock to the ISS space station, when (not quite in real-time) different services need to be coordinated in smart city development, or when sustainable foresting strategies need to be developed. In such a setting, the data management challenges are significant (cf. Fig. 4):

- To achieve the necessary speed, each partial simulation must run in the fastest native environment with ideally suited data structures, a homogeneous data format is infeasible due to the transformation effort.
- The views by human users and simulation tools show differences at various related level, in schemata, data formats, temporal and spatial context (versioning), and functionality. Fast automatically generated mappings must be defined across these views to achieve efficient propagation of changes from one view to another in the real-time setting.
- As in the earlier examples, a suitable domain-specific meta model and model-driven code generation framework must be developed to enable these functionalities.

In a recently completed dissertation, Hoppen [Hopp17] has presented a system which achieves a large part of these requirements by linking a heterogeneous set of main-memory simulation databases to a semantic exchange platform database through the mentioned four-level mapping model, where the model-driven code generation is achieved on the basis of the Eclipse Modeling Framework EMF and related toolkits.

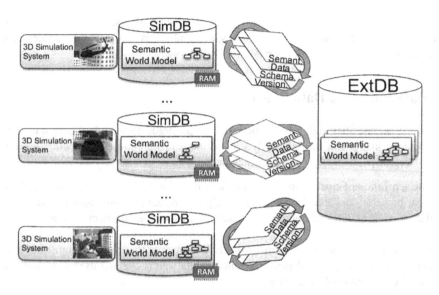

Fig. 4. Using a semantic world model database as a broker for real-time multi-level mapping between main-memory 3D simulation databases from helicopter, car, and city geography perspectives in a smart city simulation [Hopp17]

For an already operational real-world application of the system in a more limited domain-specific setting, Fig. 5 shows how this infrastructure supports a wide range of forestry tasks from multiple perspectives in several major German forest planning and management areas. Here, the central semantic model is limited to the ForestGML setting to achieve scalability and performance for real-time decision support. Many issues remain to be investigated and resolved in order to scale the approach to more

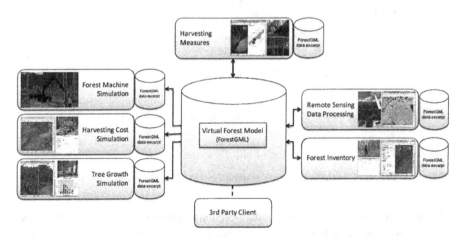

Fig. 5. Group decision and negotiation support system for sustainable forest management based on interacting 3D eRobotics simulations [Hopp17]

complex or very large systems, such as complex industrial plants. However, successful test in national and European projects in space and smart city robotics applications show the potential of such an approach.

3 Data Lakes vs. Data Spaces

For over a decade after their introduction in the mid-1990s, the standard corporate data management instrument for data analytics was the data warehouse architecture (cf. left side of Fig. 6 [JLVV03]). Key ideas included:

- The separation of current operational data from historical lightly aggregated analytic data, both for conceptual reasons and for performance reasons caused by different data models (multi-dimensional vs. relational) as well as transaction management issues,
- A structured Extract-Transform-Load (ETL) process to achieve a uniform conceptual understanding and structuring in the data warehouse and its direct basis, the so-called operational data store.

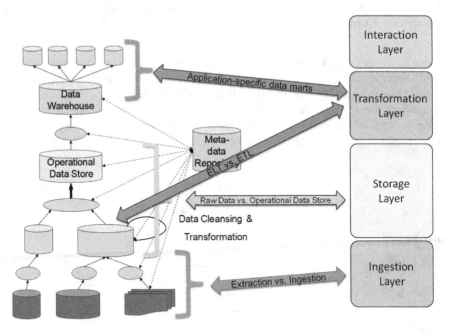

Fig. 6. Comparison of data warehouse and data lake architectures

In the mid-2000's, the original data warehouse approach came under fire from various viewpoints. The first one was a user perspective. In 2005, Franklin et al. [FHM05] noted the growing richness of media on computers and proposed that users should be enabled to create their own "data space" where a free collection of data and media objects could be managed using a network of semantic metadata. At the corporate level, managers demanded real-time decision support where current operational data should directly be added to the warehouse to immediately influence the next step – the separation of operational databases and data warehouses seemed no longer as desirable as before. For structured and semi-structured databases, main-memory databases using techniques such as column stores (e.g. SAP HANA) turned out to be a good solution, and for more heterogeneous data and media collections, but for more heterogeneous data sources, a broader architectural solution seemed necessary.

Since about 2010, therefore, the concept of data lakes has emerged [LS14] – see right-hand side of Fig. 6. Noting that the transformation step in data warehouses takes a lot of time, data lakes change the sequence from ETL to ELT – that is, data are ingested in their raw form into the data lake, and transformation is postponed much closer to the point where actual analyses are needed. This also implies that no overall data integration is necessary any more. On the other hand, the data lake systems now requires a highly complex semantic network of extremely large and extremely heterogeneous data, implying the need for new parallel algorithms and multiple kinds of frequently NoSQL data stores. This is an ongoing technical challenge in database research, which the very large social network vendors have addressed with different and often proprietary strategies, even though they do share some basic components. In this sense, data lakes can be seen as a generalization of one key aspect of the data space idea, heterogeneity and size of the data.

However, as mentioned in Sect. 1, the presence of huge data lakes with unclear data ownership and data protection also causes great political concern especially among hidden champions – small and medium companies with world-leading specialized knowhow that might get lost as a side-effect of data analytics. In other words, the second major aspect of the data space idea – that the data space was a personal thing under full control of its owner – had not transferred to the data lake idea.

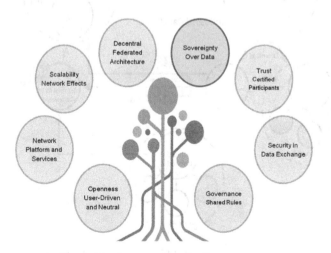

Fig. 7. Design Goals of Industrial Data Space Architecture [OL*17]

Noting this discrepancy, we (a small group of Fraunhofer institute leaders) pro-posed the idea of a so-called *Industrial Data Space* which puts data sovereignty in the center of attention. The idea was quickly taken up by several leading players in the German industrial sector and by the government which led to an initial research project aiming at a more precise specification of the design goals for such an Industrial Data Space (Fig. 7). These goals basically spell out what data sovereignty should mean in practice, and can of course only be reached by a mix of technical approaches, methods, and governance rules. It was decided to have a completely open architecture to which researchers and companies can contribute their ideas and solutions, and to focus the initial effort on the development of a Reference architecture and the set-up of an organization, called the IDS Association. A first version of the Industrial Data Space Reference Architecture [OL*17] was publicly presented at the Hannover Fair 2017, where already 26 research organizations and companies showed initial demonstrators of solutions based on aspects of the Reference Architecture. Altogether, the IDS Association has now over 70 organizational members from four continents.

Figure 8 gives an overview of the most important player roles and components. Participating companies (here: A and B) use a certified Internal IDS Connector offer or import safely containerized views on their data, with well-defined usage controls. In larger networks, specialized IDS Brokers can offer services for searching, negotiating, and monitoring service and data contracts. Other service providers can bring in data cleaning or analytics services through their App Stores, again connected to the system by IDS Connectors. For confidentiality as well as for performance reasons, service execution

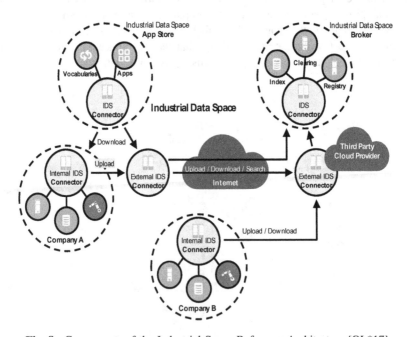

Fig. 8. Components of the Industrial Space Reference Architecture [OL*17]

should have a choice between policy-based uploading and downloading data or the service algorithms themselves which may require additional External IDS Connectors.

For a detailed description of the Reference Architecture, see [OL*17]. But this short sketch should already indicate the richness of research questions closely connected to questions of Group Decision and Negotiation, ranging from data exchange negotiations to view-based decision making integrating own and foreign data in a user-friendly manner for more informed policy and decision making. Indeed, the case study of Sect. 2.2 was already an early experiment of specializing the IDS Reference Architecture to the case of Medical Informatics and integrating specialized tools for dealing with the peculiar multi-player challenges of this field.

Acknowledgments. This work was supported in part by the BMBF, the BMWI, and several industrial partners. In particular, I would like the project management of the Industrial Data Space project (Boris Otto and Stefan Wrobel), and my co-workers Ralf Klamma for a lot of fundamental research in community IS, Christoph Quix for his contributions to data space concepts including IDS Reference Model, and Karl-Heinz Krempels for his leadership in the Mobility Broker initiative.

References

[BG*16]   Beutel, M., Gökay, M.C., Kluth, W., Krempels, K.-H., Ohler, F. Samsel, C., Terwelp, C., Wiederhold, M.: Information integration for advanced travel information systems. J. Traffic Transp. Eng. **4**, 177–185 (2016)

[BP98]   Brin, S., Page, L.: The anatomy of a large-scale hypertextual web search engine. Comput. Networks **30**(1–7), 107–117 (1998)

[BJ86]   Bui, T.X., Jarke, M.: Communications design for C-oP – a group decision support system. ACM Trans. Inform. Syst. **4**(2), 81–103 (1986)

[FJ16]   Fischer, R., Jarke, M. (eds.): Medical Data Space. Fraunhofer-Gesellschaft, Munich (2016)

[FHM05]   Franklin, M., Halevy, A., Maier, D.: From databases to dataspaces: a new abstraction for information management. ACM SIGMOD Rec. **34**(4), 27–33 (2005)

[Hopp17] Hoppen, M.: Data Management for eRobotics Applications. Doctoral Dissertation, Electrical Engineering, RWTH Aachen University (2017)

[JJS87]   Jarke, M., Jelassi, M.T., Shakun, M.F.: MEDIATOR – towards a negotiation support system. Eur. J. Oper. Res. **31**(3), 314–334 (1987)

[JLVV03] Jarke, M., Lenzerini, M., Vassiliou, Y., Vassiliadis, P.: Fundamentals of Data Warehouses. 2nd edn. Springer, Berlin (2003)

[LS14]   LaPlante, A., Sharma, B.: Architecting Data Lakes. O'Reilly Media (2014)

[NHKJ16] Neulinger, K., Hannemann, A., Klamma, R., Jarke, M.: A longitudinal study of community-oriented open source software development. In: Nurcan, S., Soffer, P., Bajec, M., Eder, J. (eds.) CAiSE 2016. LNCS, vol. 9694, pp. 509–523. Springer, Cham (2016). doi:10.1007/978-3-319-39696-5_31

[OL*17]   Otto, B., Lohmann, S., et al.: Reference Architecture Model for the Industrial Data Space. Fraunhofer-Gesellschaft, Munich (2017)

[PBMW99] Page, L., Brin, S., Motwani, R., Winograd, T.: The PageRank citation ranking – bringing order to the web. Ilpubs.stanford.de, Stanford University (1999)

[PJ16]   Pfeiffer, A., Jarke, M.: Generating business models for digitized ecosystems –
          service oriented business modeling (SoBM). In: Proceedings SmartER Europe
          Conference (2016)

[PKK15]   Petrushyna, Z., Klamma, R., Kravcik, M.: On modeling learning communities. In:
          Conole, G., Klobučar, T., Rensing, C., Konert, J., Lavoué, É. (eds.) EC-TEL 2015.
          LNCS, vol. 9307, pp. 254–267. Springer, Cham (2015). doi:10.1007/978-3-319-
          24258-3_19

[QS01]   Quix, C., Schoop, M.: DOC.COM: a framework for effective negotiation support
          in electronic market places. Comput. Networks 37(2), 153.170 (2001)

[Scha96] Schael, T.: Workflow Management for Process Organizations. LNCS, vol. 1096.
          Springer, Berlin (1996)

[WF86]   Winograd, T., Flores, F.: Understanding Computers and Cognition – A New
          Foundation for Design. Ablex Publishing Corp. Norwood, (1986)

The Strange Absence of Abstraction Levels in Designing HCI

Dov Te'eni[✉]

Tel Aviv University, 61390 Tel Aviv, Israel
teeni@post.tau.ac.il

Abstract. People process and communicate information at multiple levels of abstraction when reading, talking, solving problems, designing and interacting with computers. For example, in reading an article, actors may focus on a letter, a word, a clause, a sentence or a paragraph. At any moment, they focus on a particular level of abstraction, do something, and, under certain conditions, move back and forth to other levels until the actors achieve their goal. Not moving between levels of abstraction when necessary, decreases performance. It follows that human-computer interaction should be designed accordingly, yet there is hardly any explicit mention of abstraction levels in studies or guidelines of designing HCI. In this talk, I propose a method for incorporating abstraction levels in the design of HCI as a critical dimension of designing adaptive HCI. The talk demonstrates the ideas with examples of HCI for supporting online reading and group problem solving.

Keywords: HCI design · Feedback · Levels of abstraction

1 Introduction

The notion of levels of abstraction (LoA) has been used in numerous contexts of human information processing (Vallacher and Wegner 1987). For example, in reading this article, readers can process information at different levels such as word, sentence and paragraph. We restrict the discussion to goal directed behavior and assume that the readers have a goal they wish to accomplish, e.g., understand the article's main message or edit the article, and that they choose to concentrate on the LoA corresponding to their goal. For example, readers wishing to understand the logic of the arguments, may read the article at the level of a sentence, i.e., attempt to understand the information given in each logical step, occasionally skipping an unknown word if it does not interfere with grasping the general message of the sentence. In comparison, readers wishing to understand the article's theory and method, may read it at the level of a paragraph, i.e., attempt to grasp the message of the entire paragraph by glossing through the paragraph, seeking signs of the general message with perhaps greater attention to the beginning and ending of the paragraph or to anything in bold or italics. (Of course, we read articles in many other ways, e.g., read only its title.)

The readers' choice of a focal level of attention depends on their goal and the situation in which they pursue their goal, e.g., understand the main message of the article when

© Springer International Publishing AG 2017
M. Schoop and D.M. Kilgour (Eds.): GDN 2017, LNBIP 293, pp. 15–29, 2017.
DOI: 10.1007/978-3-319-63546-0_2

in a hurry or edit the article carefully before final approval to print. Regardless of the initial choice of a focal abstraction level, readers usually move from one LoA to another during a session of reading, e.g., they oscillate between the sentence level and the paragraph level, and at any moment, they attend to a particular focal LoA (Vallacher and Wegner 1987). These moves across levels of abstraction can be viewed as an adaptive behavior that is effective when the moves result in achieving the behavioral goals successfully.

How can designs of online articles take into account LoA? Unlike a printed article, an online article (like an e-book) affords the adaptation of the human-computer interaction (HCI) to the dynamic reading behavior of the user. The HCI can be adapted to fit the abstraction level of the user's choice or to fit an optimal abstraction level determined by the computer. In reviewing the HCI literature on designing e-books and readers, I could not find explicit consideration of LoA. Nor could I find explicit reference to LoA in the practice of designing e-books. Indeed, I found very little on LoA in design of information systems more generally, albeit, I found some limited examples of practice in certain applications that support *de facto* work at different LoA without explaining its rationale. This state of affairs is unfortunate because we are robbing ourselves the opportunity to improve future designs by considering LoA.

In the reading example, assuming that such moves between reading modes are advantageous, there are at least three broad design implications. One implication is to adapt the system so that it best serves each reading mode in turn. For instance, when scanning the article in a high-level reading mode, the reader tends to ignore less important pieces of information according to a principle called 'visual hierarchy' (Djamasbi et al. 2011), implying a design that lets the unimportant pieces almost disappear into the background. A commonly used technique to emphasize parts of the text and downplay surrounding texts is to present the page in a fisheye view. A second implication of the LoA implication is the need to design systems that support an easy transition between reading modes. For example, the system should signal clearly how to move from the current LoA to a higher or lower level and signal the arrival at the new LoA. A third, more complex, design implication of the two LoA is to prompt the user to move effectively between modes. To do so, the system has to rely on knowledge about the effectiveness of each approach for a given reading task. For instance, knowing the reader is looking for a practical solution, the system might suggest a move to a detailed-level reading when the reader gets closer to the appropriate section in the article. None of these design implications are widely implemented in online readers even though they are technically feasible, today more than ever before.

The idea of LoA is not new in systems design. Rasmussen (1986) advocated a means-ends approach to task analysis as a basis for designing systems. In more concrete and spatial aspects of HCI (i.e., working at lower levels of abstraction), the potential of interactive systems is clearer and more often utilized. An explicit consideration of LoA that I found appears in the treatment of visualization (Christoff et al. 2009). For example, interactive maps let users zoom in to a more detailed map or out to a less detailed map, which is akin to moves between LoA in more conceptual tasks. By the same token, we could expect to see systems designed to consider LoA in the treatment of abstract and symbolic tasks such as reading an article or solving a problem.

Moreover, advanced technologies are already available that can detect intentions to move to higher or lower LoA (Christoff et al. 2009) or track eye movements that can identify the user's focus of attention. Such advances in the technology will allow automatic adaptation to the desired LoA. It follows that there is a need for human-computer interaction to consider LoA in its design and that such designs are feasible technologically, and yet there is no systematic way to encourage, or even ensure, that designers incorporate LoA in their design of HCI. As these more advanced technologies become more prevalent and readily available, the desirability and feasibility will grow.

The paucity or even absence of LoA in the research and practice of HCI is puzzling. It may be a lack of awareness or the lack of methods to analyze and design LoA in HCI. Whatever the reason, not designing for LoA is a missed opportunity to improve the HCI. This talk proposes a systematic way of incorporating the idea of LoA into the design of HCI in order to achieve designs that capitalize on the idea, as demonstrated above. It begins by extending the idea of LoA in interactive systems and tying it to the design of feedback, which constitutes an important aspect of designing interactive systems. Thus, taking a critical component of system design, namely feedback, I show how designing for LoA can improve user performance by producing a better HCI. I then go beyond the idea of the LoA to suggest that the same line of thinking can be used to support the design of HCI that caters for adaptive behavior more generally.

2   Levels of Abstraction (LoA) and Feedback in HCI

2.1   LoA

An early discussion of LoA in the IS literature examined the process of solving problems in data modeling (Srinivasan and Te'eni 1995). To build an entity-relationship diagram (ERD) of a system, data modelers engage in planning, building, testing and refining, and do so at different LoA. The LoA include, from low to high LoA, properties of an entity, entities, clusters of entities and composites (e.g., a vehicle is a cluster of car and truck). This approach has been used in several studies since, e.g., Rittgen (2007) used it to study negotiation in modeling. Data modeling involves several activities such as planning and testing. What is most relevant however to our discussion is that while engaged in modeling an entity-relationship diagram of a system, data modelers adapt their modeling behavior by moving their focus of attention up and down LoA. Figure 1 shows the transitions of a participant between data-related levels of abstraction engaged in planning and testing their data model.

The figure was established based on a protocol of a think-aloud session intended to solve a data-modeling exercise to build an ERD that could enable users to answer certain queries. In their study, Srinivasan and Te'eni (1995) show that different LoA are needed to solve the problem and that moving between LoA at certain points of time enhances performance. Shifting levels can be seen as adapting to certain situations. Adaptive behavior involves working on one level, say entities, and climbing to a higher level of clusters when progress loses direction or dropping to a lower level of properties when modeling or testing fails.

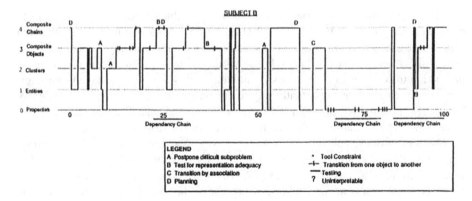

Fig. 1. Adaptive behavior in data modeling (from Srinivasan and Te'eni 1995)

Srinivasan and Te'eni (1995) explained these transitions between LoA on the basis of early work on problem solving introduced by Duncker (1945). One shift is to a higher level of abstraction, which is expected to occur when the problem solver gets lost in detail and needs the broader context to understand the details or to decide how to proceed (Duncker 1945). Another shift is to a lower level of abstraction, which is expected to occur when the problem solver feels that the higher level can no longer guide him or her on how to proceed or how to understand the problem. Importantly, Srinivasan and Te'eni (1995) showed that the transitions between LoA have consequences on performance.

From a design point of view, designs of systems that support modeling should, at the minimum, the patterns of adaptive behavior shown in Fig. 1. In this respect, these patterns represent the knowledge base required in designing for adaptive behavior in modeling.

Having identified the transitions across LoA and having characterized these shifts as adaptive user behavior, we ask what, if at all, are the consequences of certain patters of transition. In other words, why adapt and what consequences does adaptive versus non-adaptive behavior have. Knowledge of data modeling suggests that certain patterns are more effective than others in problem solving. For instance, working mostly at lower levels of abstraction was less effective than working at higher levels, although working only at higher levels of abstraction was not effective. An effective pattern was to work at the level of entities, but from time to time to climb to higher levels for short periods. There are other effective patterns, which may depend on the user's level of expertise (Srinivasan and Te'eni 1995). In any event, in this example and on the basis of knowledge about the specific phenomenon of data modeling, we can determine the relationship between patterns of LoA transitions and consequences. In other cases, we may be able to define LoA but may not be able to determine the consequences of alternative patterns.

Another application of LoA in design looked at the design of email (Te'eni and Sani-Kuperberg 2005). This work built on the notion of a focal LoA developed by Vallacher and Wegner (1987) in their theory of action identification. The theory of action identification is rooted in human information processing and claims that people represent and communicate actions at multiple levels of abstraction, and at any one moment, one of these levels may act as their focal level (Berger 1988). People tend to remain at the focal level, but shift their attention when complexity increases and breakdowns in understanding occur (Vallacher and Wegner 1987). People may shift to lower levels of abstraction when they feel that the higher level can no longer guide them as to how they should proceed or understand the message, or shift to higher levels when they get lost in detail and need a broader context for guidance. It follows that emails could be designed at multiple levels so that receivers of the message could work effectively at their current focal LoA, e.g., focus on the highest level of the message but revert to lower levels only when needed. For example, users see a message at a high level of abstraction and only when they find themselves unable to continue reading because a sentence is unclear, a more detailed presentation of the email appears (Te'eni and Sani-Kuperberg 2005). In other words, at any moment of work, the design must support the focal LoA and enable an easy transition to other LoA when necessary.

2.2 Feedback

So far, we have looked at LoA as an element of dynamic behavior and its effect on user performance. This section discusses feedback in order to examine how LoA should be taken into account when designing feedback. Feedback is arguably one of the most important aspects of HCI designs, particularly in systems that support dynamic behavior such as decision support systems (Te'eni 1992). My working assumption is that systems must be adaptive to support adaptive behavior such as moving across LoA.

Feedback has been studied extensively for years in the fields of human behavior (Annett 1969), management (Ilgen et al. 1969), decision making (Sterman 1989). In the context of human-computer interaction to support management or decision tasks, feedback can be conceived as communication between the computer and the user or as computer-mediated communication between users. As we shall see later on, both conceptualizations are relevant to design of systems that support joint or group decision-making. In either case, the first step of designing feedback is to articulate its functions and the means by which enable these functions such as choice of media and format. A primary function of feedback is to support control over the user's goal directed behavior by monitoring the direction and motivation governing the behavior.

Following the view of feedback as an element of the communication between the user and the computer, I define *feedback* as information to the user, about the user's goal-directed behavior and in response to the user's interaction with the computer system (adapted from Ilgen et al. 1979). The function of feedback is directional or motivational. Directional relates to behavioral outcomes that should be accomplished, while motivational relates to outcomes associated with rewards. Practically, feedback may affect behavior in both ways at the same time by directing behavior and incenting or reinforcing behavior (Ilgen et al. 1979). A third possible function of feedback is to support learning, which is particularly important when the relationships between the factors contributing to outcomes are complex and there is a need to develop and understanding or skill to accomplish the task (Te'eni 1992). Learning usually relies on feedback that relates not only to the outcome but also to the process leading up to the outcome, in which case it is labeled process feedback rather than outcome feedback. In the design examples discussed here, we will design outcome feedback in both directing and motivating or reinforcing behavior, as well as process feedback that explains relationships. (Process feedback, sometimes labeled cognitive feedback, relates to the process leading up to the outcome.)

However feedback is categorized or whatever function it plays, feedback is always tied to some goal the user wishes to pursue by behaving in certain ways. Goals can generally be decomposed into sub goals, each sub goal pursued with a behavior that the user attempts to control, and appropriate feedback can support the control. Furthermore, the goal decomposition defines the hierarchy of the LoA, and each LoA may therefore carry different forms of feedback.

For example, the overall goal of creating an ER diagram (ERD) can be decomposed to several sub goals such as determine an entity's properties, check that all properties are allocated to entities, and determine the relationship between two entities. Feedback at the highest level should provide information about how the ERD is progressing toward the goal of a complete and semantically correct diagram. A second feedback, feedback at a lower level, may provide for each entity a list of defined and undefined properties and present the fraction of properties already defined; this is feedback towards the goal of completing the determination of the entity's attributes. A third form of feedback could be generated while in the process of building a relationship between two entities. Say the user is in the process of selecting the type of relationship from a pull-down menu a candidate for the relationship (e.g., a one-to-many relationship), useful feedback would indicate the correctness of the relationship selected (e.g., that the relationship is infeasible given the specified properties). Each of these three forms of feedback can be associated with a particular LoA. The next section explains how the perspective of LoA informs the design of feedback.

3 Design for LoA

To recapitulate, the LOA perspective posits that people have a focal LoA at any moment of their work and that they move back and forth between LoA when necessary. Therefore, the design of feedback (and of other aspects of HCI) should adapt the design to support the focal LoA. As the focal LoA changes during the interaction, it will also be necessary to support an easy transition between levels when the user moves on her own initiative or manage and initiate transitions automatically. Using the data modeling example, we use

3.1 Fit the System to the State to Which the User Moved

Figure 2 depicts a screen that supports building and testing an ERD at different LoA, as described above. The rectangle describing the entity Person.ContactType fits the task the user is expected to perform, namely to determine and check the properties of the entity, part of which are detailed in the lower left corner (e.g. Bold is set to False). A better fit could be to move the table of properties from the lower left corner closer to the central rectangle so that it takes minimal effort to move between the two. In fact, a clever design might be able to integrate the two, in order to bring the manipulations on the entity's properties as directly as possible, e.g., when the cursor hovers over the (green) colour of the rectangle, a colour palette appears inviting the user to choose another colour.

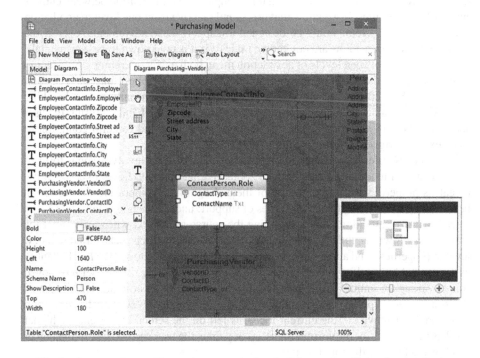

Fig. 2. A screen for working at the properties level of abstraction (Color figure online)

Designing for habitual behavior differs from designing for novel behavior that requires mindful planning and controlling. When fitting the human-computer interface to the user's expected behavior, the designer's goal is to minimize the user's effort in performing the task correctly. When fitting the human-computer interface in novel situations, the designer's goal, in addition to minimal effort and accuracy, is to support the generating of new behavior and controlling unpracticed behaviors. In the case of setting or manipulating an object's properties, behavior is assumed to be habitual (unless the user is in training). The screen is designed to allow easy manipulations of data, immediate feedback (e.g., setting the color immediately shows the color to be displayed during use under the anticipated conditions of use), and easy and therefore more accurate detection of properties and their values.

When working at a higher LoA, namely at the level of entities, the user is expected to plan the relationships between the entities. The small map of entities, when expanded, would become the centerpiece of the screen. Its graphic presentation is a better (cognitive) fit than, say, a table of relationships, because it lends itself most immediately to the way people represent relationships in their mind (Vessey and Galletta 1991).

3.2 Design Systems to Support Transitions Between LoA

The most obvious design implication, often ignored in practice, is to enable easy transitions between LoA. This would mean, for example, an easy transition between entities and clusters of entities. If we know that users tend to move from one level to another on a dimension, the system should be designed to support these movements, even if the consequences of transitions are not known (at least if it is not known to be detrimental). Easy transitions between levels of abstraction mean both easily operating the system to reveal and move the focus of attention to another level and supporting the transition cognitively (Sun 2012). The move from one level to another usually means breaking away from the habitual behavior into which the user has settled and requiring the user actively think how to proceed, and this requires some new and forceful condition or some external trigger (Louis and Sutton 1991).

Technically effecting a transition to another level usually involves some form of direct manipulation, like a single click to zoom in or an option to hover over an entity, reveal its properties and move to one of them. Cognitively supporting the transition between levels includes at least two types of support: underscoring the new state as the current focus of attention, and maintaining the source as context when arriving at the target. In other words, the user often needs to realize that the focus has shifted to a new level of properties. And at the same, he or she needs to take time to see the level of entity as the context for working on the properties. For example, in Fig. 2, the user has moved from working on entities to focusing on the properties of a particular entity labeled 'Person.ContactType'. The small map of entities (the higher level of abstraction from which the user moved to the current focus) is left on the screen to present the higher level as the context for the current focus. At any time, the user can go back to the higher level by clicking on the small map.

3.3 Design Systems that Guide Advantageous Behavior

The last of the five steps of the procedure for designing adaptive behavior systems requires knowledge of performance measures in the particular domain of the user's work. Furthermore, this step is only feasible if there is a convincing argument for affecting consequences advantageously by manipulating the human-computer interaction (step #2 above). The research quoted above argues that certain patterns of transition are more effective than others. For example, remaining too long at the lowest levels of abstraction without occasionally taking a more comprehensive view by climbing to a higher level of abstraction will cause people to make errors. The system can detect fairly long periods of working on properties by monitoring the user's manual inputs to the system (in other cases, the system could monitor the user's gaze). The system can then alert, suggest, or force corrective action like moving to a higher level. In other cases, it may be important to consider not only the proportionate time spent at different levels, but also the sequence of activities.

Technologies that monitor users' behavior and conditions are becoming widespread in some areas like health and safety at work, but will most likely spread to many other domains of life. The nearly constant accessibility to devices like the cell phone, online watch and wearable devices in general, as well as knowledge about what is effective in which conditions makes it feasible to guide advantageous behavior. A simple example is RunKeeper's (an App to plan and monitor jogging activities) real time health graph, which could easily be supplemented with alerts of when you should accelerate to meet your running goal, or slow down to maintain your health. Similarly, as well as signaling when the user should move to another level of abstraction, a modeling system could motivate the user with graphical depictions of successful patterns vis. a vis. her own actual pattern.

4 Design of Feedback in Group Decision-Making

4.1 Task, LoA and Feedback

Feedback in computerized systems that support joint (or group) decision-making can be conceived within two contexts of communication: communication between the computer and the user (as was the feedback described in the data-modeling example above) and communication between users mediated by the system. This section discusses feedback in the context of a personnel-screening task in which a team of recruiters wishes to screen candidates. The recruiters initially work alone to form their individual judgments of the candidates but eventually are exposed to the judgments of their fellow recruiters. As a final step, the recruiters work jointly to negotiate a collective ranking. As demonstrated below, communication between recruiters provides opportunities for interpersonal feedback between colleagues in addition to feedback generated by the computer as part of the HCI.

Sengupta and Te'eni (1993) experimented with this setting to test the impact of feedback on the decision maker's cognitive control. They created teams of three recruiters that ranked applicants according to structured profiles composed of three attributes: work experience, test scores, and education. To avoid biased judgments, the recruiters did not receive other information, such as photos, gender or age. Each of the three attributes was presented to the recruiters as an integer evaluation (on a 1–9 scale). The recruiters (users) worked individually on desktop computers but were linked through a network so that they received information about their colleague's judgments as they progressed with their own judgments.

The recruiter's first goal was to evaluate ten candidates. The evaluations from the three recruiters would later serve as a basis for a collective ranking of the ten candidates. The user (recruiter) worked at two broad LoA: setting the weights for to the three attributes, which applies to the entire set of candidates, and evaluating each candidate. The system supports both levels with appropriate feedback. The user can begin by setting the weights and receiving feedback that visually reflects the relative importance of the three attributes and compares the current weights to previous weights and to weights set by others. The user can then inspect each applicant's profile and rate the applicant overall (also on a 1–9 scale), receiving feedback computed according to a theoretical model of social judgment (Hammond et al. 1980). At the user's discretion, feedback from fellow recruiters showing their ratings can be shown side by side with the user's rating so that the user has the opportunity to adapt her or his own behavior by changing ratings.

4.2   Fitting the System to the Focal LoA

Figure 3 shows the screen for working alone (the user is John) at both LoA: determining the overall strategy (setting the weights) and rating an individual candidate. How does the screen design adapt to a change in the user's focal LoA? When the user moves to setting weights, the working area that includes the relevant feedback is highlighted and the rest of screen is dimmed. The relevant feedback includes the newly inputted weights in comparison to previous weights as well as cognitive (process) feedback showing visually on the horizontal bar chart, how each bar stretches or shrinks as each weight is changed. Moreover, the consistency of applying the strategy is also shown. The theory-based index of consistency is feedback about the implications of the user's action in setting the weights. Thus, the entire set of feedback signals are designed to hold the user's attention to the focal LoA.

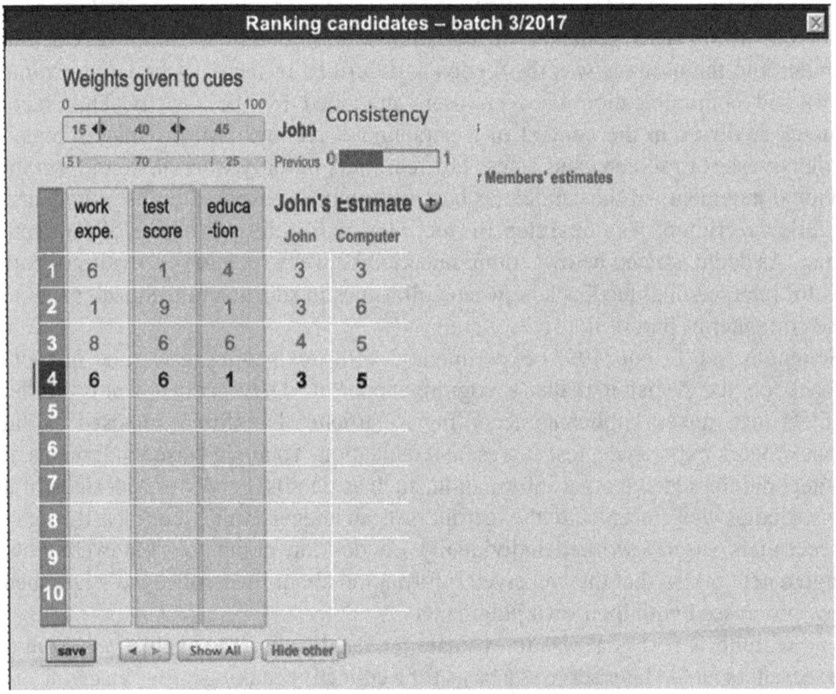

Fig. 3. Working alone screen for ranking candidates

Similarly, when the user moves to work at the level of a single candidate, the corresponding feedback must be at the focus of attention. Say the user John is considering the values of the three attributes in the highlighted row of the fourth candidate. The values (6, 6, 1) for the work experience, test scores and education, respectively), feedback showing the predicted evaluation appears under 'computer. In other words, after the user inputs an evaluation (3), the user's predicted evaluation (5) is feedback generated by the computer. The user can always examine the feedback shown at the higher level to guide individual evaluation, which is essentially a move up the LoA to see the context of the lower LoA.

Once the user elects to communicate with the two other recruiters, their input and predicted values are displayed in the adjacent columns, as shown in Fig. 4. This is feedback that comes from others but is relevant to the user's action. This feedback at the level of evaluating a single candidate. Feedback involving other recruiters is also generated for the higher LoA that includes a match between pairs of recruiters, their consistency and a comparison of their weights with to the user's weights.

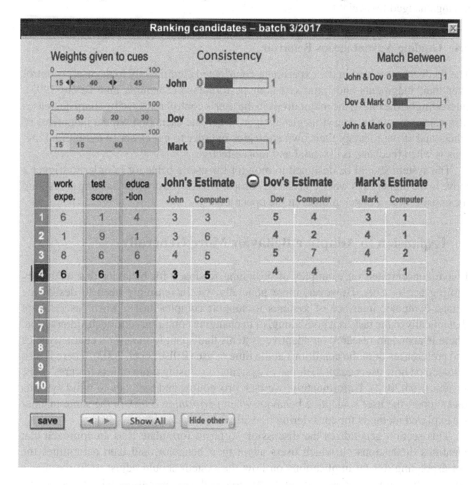

Fig. 4. Working with others in ranking candidates

4.3   Supporting Transitions Between Levels

In this case, the transitions between LoA are relatively easy because the upper and lower parts of the same screen are respectively the higher and lower LoA. The awareness of the two levels is high and the affordance of the system for shifting LoA is obvious – all the user has to do is to move the cursor to the area of interest. In contrast, the shift (toggle) from working alone to working with others requires pressing the $\pm$ or the buttons at the bottom of the screen. Beyond the physical motion to shift levels, the two levels are syntactically similar, using the same colors for attributes, as well as the order of recruiters. The formatting decisions help the user move from one level to another by minimizing cognitive effort in the move. The logical relationship between levels is further clarified by showing the change in predicted values when changing the user's weights (this happens when the user moves from the upper area towards the lower area, having changed the weights).

4.4 Guiding Advantageous Behavior

The feedback provided in the experiment contributed significantly to the users' control over their judgments and improved their performance (Sengupta and Te'eni 1993). Interestingly, the one of the major drops in the user's control reflected by sharp decreases in consistency of applying strategies happens when the users first see the judgments of others and strive to adapt their own strategies in order to get closer to a joint ranking. This is when feedback is essential and most effective.

The system should be designed to monitor the user's behavior in order to trigger a move to the higher LoA when consistency drops. This can be done by flashing or resizing the consistency or match indices in the upper area.

5 Expansion to Adaptive Behavior More Generally

I have concentrated on examples of designing feedback for behavior that adapts by shifting across LoA. However, more generally, we increasingly need to design the human-computer interface of systems to support complex tasks, where users adapt dynamically to the task as it goes along, or to changing conditions during the interaction. There is abundant research on adaptive systems that adapt to the user's characteristics and preferences or to the conditions at the time of use (Billsus et al. 2002). There is also research on how users adapt their use of systems, e.g., using more or less features of the system (Sun 2012). Unfortunately, we lack procedures and methods to build systems that support the user's adaptive behavior when *performing a task* in the same manner we explored methods for considering transitions in LoA.

This section generalizes the discussion so far to formulate it as an approach that identifies dimensions on which users adapt their behavior, and then determines the corresponding design implications on how the system should adjust to fit the user's adaptive behavior. As noted above, technologies that supply users with physical and physiological data such as location or blood pressure to help them adapt their physical behavior are commonplace. Our concern is with mental tasks like reading or solving problems that rely on abstract feedback in order to adjust behavior. Future work will examine technical solutions based on advanced technologies for data acquisition and analytics to supply appropriate feedback but for now, I generalize our discussion assuming current technologies.

I propose a procedure of five steps to design systems that support adaptive behavior (regarding moving across LoA as one case of adaptive behavior):

1. Identify dimensions for adaptive behavior in a given activity
2. Determine the consequences of transitions between states on a dimension
3. Design systems to support transitions between states
4. Fit the system to the state to which the user moved
5. Design systems that guide advantageous behavior

For example, adaptive behavior can be transitions between a mode of automatic behavior, when things are familiar, to controlled behavior, when things become strange or unfamiliar (Louis and Sutton 1991).

Our examples have looked at how to support an individual's adaptive behavior by achieving better cognitive fit within modes and optimizing moves between modes. We can use the same approach with other aspects of supporting adaptive behavior, beyond individual cognition. For instance, nowadays modeling involves working in different modes: alone, in intimate groups and in large groups. Applying the five-step approach and relying on knowledge of the effectiveness of the different modes, will change the design of many enterprise wide systems. Another telling example is the impact of adapting information about the user (personalization) according to the user's dietary behavior with a system that shows the effect of food consumption on the user's health (Ronen and Te'eni 2013). The study compared personalized versus generalized information with respect to the users' attitudes and behavior. Subjects receiving the personalized feedback reported a more positive attitude and a greater propensity to follow the doctor's recommendations in order to improve their well-being. In other words, a good fit not only supports adaptive behavior but also shapes it by affecting attitudes.

In the future, adaptive behavior may be supported so that the fit is affective, as well as cognitive and physiological. This would require monitoring emotions, which although not an easy task is certainly feasible. Similarly, adaptive behavior must be supported in social settings too.

The technological feasibility of supporting adaptive behavior in human-computer interaction is increasing dramatically, and so will the expectations of adaptive systems. As demonstrated above, in order to proactively recommend or even trigger adaptive behavior, the system must recognize the user's current position on dimensions of change and the current conditions. More and more sophisticated technologies (e.g., the Internet of Things) will be available to detect information about the user, the task and the setting. For instance, a user engaged in collaborating with remote others could be notified of the susceptibility of miscommunication at the receiver's end on the basis of sensors detecting noise in the communication. The system can then adapt the conversation appropriately.

More sensors transmit more information to users about the environmental conditions as well as the bodily conditions; most probably, people will feel pressurized to adapt accordingly. The new information technologies and infrastructures need, but also enable, more adaptation. The five-step procedure for designing systems that support adaptive behavior relies on the knowledge of the user, task and setting. The first step is to use the knowledge of the user's behavior when solving the task in order to determine the dimensions of changing behavior, therefore, more knowledge and more sophisticated computing and data analytics will facilitate better support for adaptive behavior.

References

Annett, J.: Feedback and human behaviour: The effects of knowledge of results, incentives and reinforcement on learning and performance (1969)

Billsus, D., Brunk, C.A., Evans, C., Gladish, B., Pazzani, M.: Adaptive interfaces for ubiquitous web access. Commun. ACM **45**(5), 34–38 (2002)

Christoff, K., Keramatian, K., Gordon, A.M., Smith, R., Mädler, B.: Prefrontal organization of cognitive control according to levels of abstraction. Brain Res. **1286**, 94–105 (2009)

Djamasbi, S., Siegel, M., Tullis, T.: Visual hierarchy and viewing behavior: an eye tracking study. In: Jacko, Julie A. (ed.) HCI 2011. LNCS, vol. 6761, pp. 331–340. Springer, Heidelberg (2011). doi:10.1007/978-3-642-21602-2_36

Feigh, K.M., Dorneich, M.C., Hayes, C.C.: Toward a characterization of adaptive systems a framework for researchers and system designers. Hum. Factors J. Hum. Factors Ergon. Soc. **54**(6), 1008–1024 (2012)

Hammond, K.R., McClelland, G.H., Mumpower, J.: Human Judgment and Decision Making: Theories, Methods, and Procedures. Praeger Publishers, New York (1980)

Ilgen, D.R., Fisher, C.D., Taylor, M.S.: Consequences of individual feedback on behavior in organizations. J. Appl. Psychol. **64**(4), 349 (1979)

Kennedy, M., Te'eni, D., Treleavan, J.: Impacts of decision task, data and display on strategies for extracting information. Int. J. Hum. Comput. Stud. **48**, 159–180 (1998)

Louis, M.R., Sutton, R.I.: Switching cognitive gears: from habits of mind to active thinking. Hum. Relat. **44**(1), 55–76 (1991)

Rasmussen, J.: Information processing and human-machine interaction. An approach to cognitive engineering (1986)

Rittgen, P.: Negotiating models. In: Proceedings of the 19th International Conference on Advanced Information Systems Engineering, pp. 561–573. Springer, June 2007

Ronen, H., Te'eni, D.: The impact of HCI design on health behavior: the case for visual, interactive, personalized (VIP) feedback. In: ICIS 2013, Milan, 16–18 December, 2013 (2013)

Sengupta, K., Te'eni, D.: Cognitive feedback in GDSS: improving control and convergence. MIS Q., 87–113 (1993)

Srinivasan, A., Te'eni, D.: Modeling as constrained problem solving: An empirical study of the data modeling process. Manage. Sci. **41**(3), 419–434 (1995)

Sterman, J.D.: Modeling managerial behavior: misperceptions of feedback in a dynamic decision making experiment. Manage. Sci. **35**(3), 321–339 (1989)

Sun, H.: Understanding user revisions when using information system features: adaptive system use and triggers. MIS Q. **36**(2), 453–478 (2012)

Te'eni, D.: Analysis and design of process feedback in information systems: Old and new wine in new bottles. Account. Manag. Inform. Technol. **2**(1), 1–18 (1992)

Te'eni, D.: Designs that fit: an overview of fit conceptualizations in HCI. In: Zhang, P., Galletta, D. (eds.) Human-Computer Interaction and Management Information Systems: Foundations, pp. 205–221. M.E. Sharpe, Armonk (2006)

Te'eni, D., Sani-Kuperberg, Z.: Levels of abstraction in designs of human–computer interaction: the case of e-mail. Comput. Hum. Behav. **21**(5), 817–830 (2005)

Timpf, S., Volta, G.S., Pollock, D.W., Egenhofer, M.J.: A conceptual model of wayfinding using multiple levels of abstraction. In: Frank, A.U., Campari, I., Formentini, U. (eds.) GIS 1992. LNCS, vol. 639, pp. 348–367. Springer, Heidelberg (1992). doi:10.1007/3-540-55966-3_21

Vallacher, R.R., Wegner, D.M.: What do people think they're doing? Action identification and human behavior. Psychol. Rev. **94**(1), 3 (1987)

Vessey, I., Galletta, D.: Cognitive fit: an empirical study of information acquisition. Inform. Syst. Res. **2**(1), 63–84 (1991)

General Topics in Group Decision and Negotiation

University of Hohenheim

Stream Introduction: General Topics in Group Decision and Negotiation

Mareike Schoop and Philipp Melzer

Chair of Information Systems I, University of Hohenheim, Schwerzstr. 40, 70599
Stuttgart, Germany
{schoop,melzer}@uni-hohenheim.de

1 Introduction

The stream "General Topics in Group Decision and Negotiation" represents the broad range of this research area. Group decision and negotiation are intertwined yet distinct research areas. Negotiations are iterative processes involving at least two participants who experience conflicts of interest. To resolve the conflicts, they communicate and make decisions according to their individual preferences. The overall goal is twofold, namely ensuring mutual understanding and reaching an agreement as a compromise decision for the participants. Research topics include theories, models, and systems supporting group decisions and negotiations.

This volume incorporates the best three full papers submitted to the stream "General Topics in Group Decision and Negotiation". Published articles specifically deal with the topics of (1) collaborative processes, and (2) game-theoretic approaches in group decision and negotiation.

2 Overview

Investigating collaborative processes in group decisions and negotiations *Steven Way and Yufei Yuan* address the need for more effective collaborative disaster response systems developing a framework for context-aware multi-party coordination systems based on a grounded theory approach.

Focusing more on the game-theoretic approaches in group decision and negotiation *Takahiro Suzuki's and Masahide Horita's* paper investigates convergent menus of social choice rules dwelling on social choice theory. *Antonio Jimenez-Martin, Hugo Salas, Danyl Perez-Sanchez, and Alfonso Mateos* analyse the remediation of the Zapadnoe uranium mill tailings site from the perspective of group decision making.

A Framework for Collaborative Disaster Response: A Grounded Theory Approach

Steven Way[1](✉) and Yufei Yuan[2]

[1] Mohawk College, Hamilton, Canada
steven.way@mohawkcollege.ca
[2] McMaster University, Hamilton, Canada
yuanyuf@mcmaster.ca

Abstract. Our society faces many natural and man-made disasters which can have severe impact in terms of deaths, injuries, monetary losses, psychological distress, and economic effects. Society needs to find ways to prevent or reduce the negative impact of these disasters as much as possible. Information systems have been used to assist emergency response to a certain degree in some cases. However, there continues to be a need to develop more effective collaborative disaster response systems. To identify the core features of such systems, a grounded theory research method is used for data collection and analysis. Data from firsthand interviews and observations was combined with literature and analyzed to discover several emergent issues and concepts regarding collaborative disaster response. The issues and concepts were organized into four categories: (i) context-awareness; (ii) multiparty relationships; (iii) task-based coordination; and (iv) information technology support, which together identified the needs of collaborative disaster response coordination. Using evidence from the data, these factors were related to one another to develop a framework for context-aware multi-party coordination systems. This study contributes to the field of emergency management as the framework represents a comprehensive theory for disaster response coordination that can guide future research on disaster management.

Keywords: Disaster response system · Grounded theory · Context-awareness · Multiparty coordination · Requirements analysis

1 Introduction

Large scale disasters such as earthquakes and terrorist attacks often occur unexpectedly and cause significant damage to our society. However, as evident in past disasters, we are often not well prepared to deal with possible disasters and collaborative responses have been poorly organized [1]. Disaster response faces multiple challenges such as "great uncertainty; sudden and unexpected events; the risk of possible mass casualties; high amounts of time pressure and urgency; severe resource shortages; large-scale impact and damage; and the disruption of infrastructure support necessary for coordination like electricity, telecommunications, and transportation. This is complicated by factors such as infrastructure interdependencies; multi-authority and massive personal

© Springer International Publishing AG 2017
M. Schoop and D.M. Kilgour (Eds.): GDN 2017, LNBIP 293, pp. 33–46, 2017.
DOI: 10.1007/978-3-319-63546-0_3

involvement; conflicts of interest; and the high demand for timely information [2]."
Many of the identified challenges influence the ability or desire of responders to collaborate, hence, these challenges can greatly affect the overall effectiveness of the
response effort [3].

The use of information technology has been shown to improve the capacity of
disaster response by improving the ability of emergency management agencies to
coordinate complex intergovernmental systems [4]. In addition, several types of
emergency response systems have been proposed and deployed in the past to improve
various disaster response and coordination problems [5–8]. There continues to be
importance and planning applied to disaster response. However, there is still much to
improve as many of the existing theories or solutions were developed prior to major
advances in technology which have rendered some elements of solutions obsolete, or
are missing some elements from the theory due to unforeseen advances in technology.
For instance, wireless communication, social media and social networks are increasingly used to support information gathering and coordination, while satellite and drone
technology helps to gather information about disaster situations.

With recent advances in information and communication technology, can we better
utilize such information systems to support disaster response, improve coordination,
and reduce a disaster's negative impact on society? What are the characteristics that
future information systems should possess for improved disaster response and coordination? In this paper, we will follow a grounded theory approach to collect data and
analyze the general requirements and principles of collaborative disaster response
systems.

1.1 Emergency Management Background

"Emergency management has ancient roots. Early hieroglyphics depict cavemen trying
to deal with disasters.... As long as there have been disasters, individuals and communities have tried to do something about them; however, organized attempts at
dealing with disasters did not occur until much later in modern history [9]". Emergency
management was primarily perceived as a function of law enforcement and fire personnel [10]. Over time, arguments were made that emergency management should
adopt risk management principles [11]. In doing so, emergency management became a
greater priority for public administrators in the form of risk management which has
been defined as the systematic application of policies, procedures, and practices to the
tasks of identifying, analyzing, assessing, treating, and monitoring risk [12]. The
Prevention-Preparedness-Response-Recovery (PPRR) model for emergency management is considered part of the risk management approach in the treatment of risks [13],
and is the model with which most responders are familiar.

1.2 Integrating Information Systems

As technology takes on a more prominent role in society, a few frameworks have been
proposed to aid in the understanding of emergency management information concepts

and systems. For example, Turoff et al. [7] used nine premises to make design recommendations for a dynamic emergency response management information system (DERMIS). Yuan and Detlor [14] examined the major task requirements and associated key issues for intelligent mobile crisis response systems. Abrahamsson et al. [15] identified four major challenges to the analysis and evaluation of emergency response systems. Van de Walle and Turoff [16] discussed many recent advances and challenges in both individual and group decision support systems (DSS) for emergency situations. Other research examined the many challenges in dealing with command and control mechanisms, conflict of interest management, authorized territory, and inter-agency communication barriers [1]. Chen et al. [3] developed a framework to analyze coordination patterns occurring in the emergency response life cycle. More recently, Yang et al. [8] developed design principles for an integrated information platform for emergency responses based on a case study of the 2008 Beijing Olympic Games. However, Franco et al. [17] addressed the limitations to much of the research being produced in the Information Systems in Crisis Response and Management (ISCRAM) community: "the lack of a deep multidisciplinary dialogue about what constitutes scientific evidence, the domination of case study methodology and the lack of alternative methods to build the confidence in causal and generalizability claims, and little effort to analytically or inferentially generalize from the findings offered in the proceedings to a theory of disaster management." In this study, we attempt to develop a theory based on scientific evidence that overcomes the limitations identified by Franco et al. [17] in traditional ISCRAM studies.

2 Research Method

2.1 Methodology Selection

This study is exploratory and seeks to answer general questions such as how and why type questions in a complex disaster response context involving multiple aspects of social, organizational and technical issues. Together, the exploratory nature and the contextual information to be studied indicate a qualitative based research methodology is preferred.

After considering several qualitative approaches, it was determined that grounded theory is the most suitable methodology for this study as it is both exploratory in nature and ideally suited for theory development. While design science was considered, it was ruled out because our main objective is to identify basic design principles rather than to create system artifacts from this study. This study can lay the foundation for design science research. Furthermore, the substantive area of research is new based on the adoption of a comprehensive collaborative perspective to disaster response. It was concluded that grounded theory's methodology characteristics for selection fit the desired output of the study. Grounded theory methodology can be summarized as an iterative and integrated process of data collection, analysis, coding, and conceptualization which culminates in theory generation. Figure 1 provides a visual overview of the methodology used in this study. For more detailed discussion on Grounded Theory

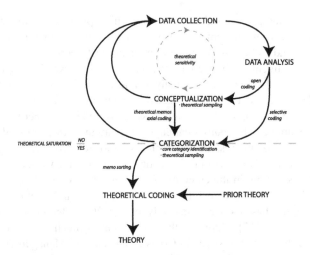

Fig. 1. Visualization of the grounded theory method used

methods in Information Systems research please refer to [18] and a special issue on Grounded Theory in European Journal of Information Systems [19].

2.2   Data Collection

Data Collection Approach
With grounded theory studies, everything is considered data and the researcher is encouraged to collect rich versatile forms of data [20–22]. For this study data collection was designed to capture the apparently complex disaster response process in order to deliver a comprehensive perspective on disaster response. To achieve this, data sources were selected to study response issues that exist in regular emergency responses which then were compared to response issues in larger disasters. The reasoning for comparing both emergency response and disaster response data was to capture the requirements that occurred during the transition from an emergency to disaster.

Data collection took place in two phases. The first phase was the collection of primary data using in-person interviews, direct observations of responders in the field, and direct observations of responders performing multi-agency training exercises. The second phase of data collection for large disaster response used alternative sources of secondary data previously gathered by other researchers or media reporters including prior academic studies, government investigations, media sources, and documentaries. The two phases of data collection provided very different but relevant sources of data for use. Primary data collection permitted interaction and follow-up lines of questioning in a semi-structured manner that were not available with secondary data sources. Furthermore, direct observation provided additional insight and triggered additional questions arising from researcher experiences in the operating environment that were relevant but may not have otherwise been studied during secondary data collection. Secondary data collection was used to provide additional samples from data collection to address any experience gaps from primary data collection that arose from participants with limited participation in larger scale disasters. Secondary data collection also provided theoretical saturation for a general theory with the inclusion of data collected from disasters which occurred in different parts of the world where different operating conditions exist. Together, the two sources of data collected in different phases provided complementary data that supported the emergent theory in this study.

Study participants for firsthand data sources were chosen from the typical set of first responders including Fire Services, Emergency Medical Services, and Police Services as well as agencies included a ministry of health, a regional fire response authority, and a regional emergency management authority. Semi-structured interviews were used to collect their experience and opinion on emergence response. In addition to interviews, data were also collected from observing several shifts for police and EMS at communication centers as well as at actual emergence handling work sites and multi-agency training exercises. Data were also collected from secondary sources including newspaper articles, internet reports, books related to disasters, documentaries, and quick response research reports, and government reports. The main secondhand data sources that emerged from the search process were Quick Response Reports from the Natural Hazards Center of the University of Colorado at Boulder (http://www.colorado.edu/hazards/research/qr/), and reports from the United States Government Accountability Office (GAO) (http://www.gao.gov/). These two sources were mainly used for the secondhand data sources as they are two existing repositories of high quality data covering several disasters over several years in several locations.

2.3 Conceptualization and Analysis Process

The conceptualization process involved merging, clarifying, splitting, and eventually categorizing the concepts that were represented in the data. The purpose of this process was to define distinct individual concepts that had no overlap with other concepts yet were fully represented and supported in the data. Using NVivo, documents were analyzed and coded for concepts. Concepts were passages of data or key observations that indicated or stated the actions performed during response. In addition, any motivational information expressed by responders for why actions were performed was also coded. Any perceived explanations for the motivations of responders were tracked in the creation of memos. Memos also tracked any perceived relationships that emerged amongst concepts.

Figure 2 is an example from a document studied which illustrates how stored data in NVivo software were analyzed and NVivo software tools were used to highlight examples and code the concepts they represent. Authors' opinions were not coded so as not to bias interpretation of the open codes. However, opinions influenced the development of theoretical memos.

Table 1 demonstrates the thought processes that occur during the development of theoretical memos. Memos continue the analysis process by abstracting relationships amongst codes to a higher level. In theoretical memos, open codes are organized and

PROGRESS

When compared to the findings of earlier studies, this evaluation of the response to Hurricane Georges shows evidence of improved relief operations. Areas witnessing progress include an immediate declaration of the disaster, the distribution of appropriate and usable aid, a higher degree of coordination among humanitarian actors, increased experience and training of relief workers, and the integration of humanitarian assistance and development. These issues will now be discussed in order.

Immediate Declaration of the Disaster

This investigation reveals that the Dominican government quickly declared a state of emergency after the disaster and likewise established a curfew that same night to prevent any social disturbances that might have arisen.49 Respondents also noted that numerous government agencies started to work right away to clean up the debris that Georges left and to provide relief to its numerous victims.50 Furthermore, those interviewed observed that political officials did not delay in requesting assistance from the international humanitarian community.51 Thus, the government did acknowledge the disaster as well as its need for help from outside sources.

The Distribution of Appropriate and Usable Aid

Findings about the nature and provision of aid in this research project are not conclusive as many of the respondents had no comment on such questions or were not aware of any difficulties in this area.52 It appears on the surface though - in spite of minor and normal problems that could arise in any large relief operation - that donors and relief providers were "much more conscious" about what they were giving to the victims of Hurricane Georges.53 Four examples support this view. First, while a few of those interviewed asserted that there was too much clothing being provided,54 the greater number did not mention an excess of any other particular type of aid.55 A Program Manager for the United Nations even doubted that an overabundance of relief was possible.56 Second, although there was one reported case of diet medicine showing up in relief supplies,57 there was no additional evidence of aid being sent which was not requested. This is probably due to the fact that non-governmental organizations relay pertinent information to prospective donors.58 Third, and despite the fact that a truck delivered contaminated water to a shelter,59 there were no further reports of unusable aid. In fact, some respondents were impressed by the quality and condition of the aid that was arriving.60 Forth, there was agreement among those interviewed that the aid was appropriate for the disaster context. This could have been due to the fact that donors attempt to communicate frequently with victims and their representatives in disaster areas,61 or also because some non-governmental organizations receive money from international donors and are able to buy the necessary goods and supplies locally.62 Only one respondent replied that he had seen relief that was inappropriate for the disaster context (i.e. winter coats in a tropical climate).63 Therefore, it appears that aid was generally beneficial to the victims of Hurricane Georges.

A Higher Degree of Coordination

The low level of collaboration among various agencies and organizations has long been a criticism of relief operations, and the respondents' view of Hurricane Georges in this study was not significantly different. For instance, a Program Development Specialist for USAID stated that everyone in the public, non-governmental and private arenas was doing their own assessments of the disaster.64 A Red Cross official stated that the Civil Defense did not advise them of where the shelters were going to be located.65 Also, a respondent stated that some organizations were working alone in various parts of the country.66 Yet the interviews of this study also indicated that coordination was a significant feature of the relief operation after Hurricane Georges. As an example, officials from foreign nations worked closely with the Dominican government to help fulfill victims' needs.67 Non-governmental organizations in the Dominican Republic interacted with other domestic and international disaster relief agencies.68 Local social groups and other humanitarian organizations were in constant contact with emergency managers in the Dominican Republic.69 Government officials received assistance from businesses in the private sector.70 Churches consulted with the Civil Defense and non-governmental organizations.71 Finally, churches, humanitarian agencies, and governments were exchanging information and assistance with counterparts and/or various branches of their respective organizations.72 It is probable that coordination was more prevalent in the response to Hurricane Georges than in the relief operations of 20 years ago. Respondents felt for the most part that "it is impossible to work without collaboration" as coordination facilitates the sharing of resources (i.e. information and supplies) and minimizes the duplication of effort.73

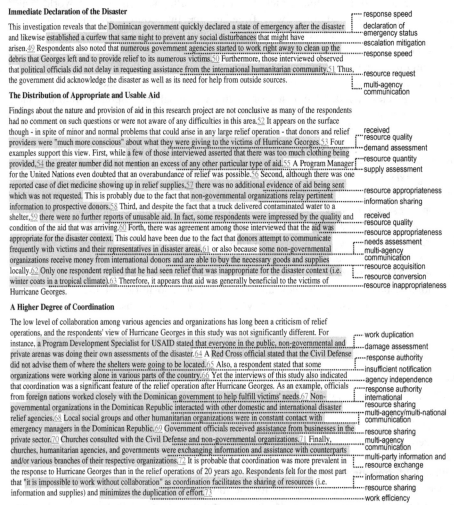

Fig. 2. Open coding sample excerpt from a QRR report [23]

grouped into concepts. Any perceived relationships between concepts are explored and used to inform future data gathering processes.

In time, themes or categories may emerge from the relationships amongst concepts and these are also explored using memos. This is shown in Table 2, which was generated from the analysis of Table 1. Eventually, a core category emerges which is the major theme represented in the data for the phenomenon. In this study, multi-party coordination emerged as the major theme when examining disaster response. The last row in Table 2 shows how coordination emerged as the core category from the conceptualization of the sample in Fig. 2. It was supported by the sub-themes that emerged in the memos.

Table 1. Theoretical memo sample from Fig. 2 excerpt [23]

Concept	Open code	Memo
Declaration of Emergency Status	• Declaration of Emergency Status	Emergency status appears to play a very important role in emergency response. Its declaration triggers different protocols and response behaviors. It informs the public that special rules are now in effect. It is also a required condition for some international organizations to provide resource assistance. Without an emergency declaration, some agencies will not be able to assist
Damage and Victim Needs Assessment	• Needs Assessment • Damage Assessment	The importance of identifying the damage to society and needs of society appears to be very important to disaster response. The assessment information is used to determine the level of response and the type of resources required to help victims and prevent the escalation of a disaster
Response Authority	• Response Authority • Agency Independence	Response authority was often implied. Jurisdictional authority to the government was implied due to the interaction with foreign nations. Some agencies chose to act independently in some locations, while in other locations there was an implied dependence on authority to provide information and guidance

Table 2. Exploring relationships sample from Fig. 2 excerpt [23]

Themes	Conceptual relationships	Memo
Contextual information	• Damage and Victim Needs Assessment • Resource Availability Assessment • Declaration of Emergency Status	Damage and victim needs, resource availability, and emergency status are all pieces of contextual information that responders need when planning a response. This information affects how tasks go about getting accomplished if protocols change due to the emergency status. It also affects what agencies will need to be involved in the response, and what resources will be required, and if there are enough resources to meet the needs of the response

(continued)

Table 2. (*continued*)

Themes	Conceptual relationships	Memo
Multi-agency relationships	• Multi-Agency Communication • Notification • Response Authority	Communication amongst agencies and the notification of agencies to situations and plans seems to affect overall disaster response. These activities would seem to be controlled by any established hierarchy or response authority amongst responding agencies. Grouping these concepts into a category called multi-agency relationships would address their common elements. Relationships establish how the agencies communicate with one another, who needs to be informed, and what information should be shared with different stakeholders as directed by those in authority or through an established response structure
Task management	• Resource Management • Task Performance	Managing the resources required to respond to a disaster, and the effectiveness in which tasks are performed appear to be related to overall task management. The ability to act quickly without the proper resources might cause problems with the response, whereas quick action with proper resources can be seen to improve response or mitigate damage. Task management incorporates the performance of activities and the management of resources in order to complete tasks
Coordination	• Contextual Information • Multi-Agency Relationships • Task Management	A higher level relationship appears to be emerging where the integrated management of information, relationships, and activities are responsible for overall disaster response effectiveness. Coordinating the exchange of information amongst agencies for planning responses and the execution of tasks with proper resources appears to enable more effective disaster response

2.4   Emergent Categories and Concepts

The four categories that emerged are Context-Awareness, Multi-Agency Relationships, Task Management, and Emergency Support Technology. Figure 3 provides a graphical illustration of the core category, factors, and supporting concepts from the data. The concepts which influence these factors are ordered with respect to their apparent priority to responders as indicated by the prevalence of the concepts present in the response data. Border thickness around the concepts also reflects a relative importance (measured by the frequency of concept occurrence) as present in the data.

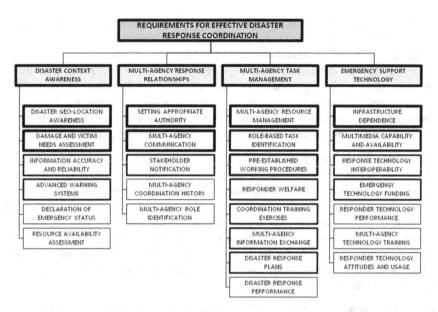

Fig. 3. Emergent concepts and their relative prevalence in data

There is a total of 26 important concepts identified in four categories. These concepts emerged from the stories and information provided by participants and literature reviews. They may also be considered requirements since they were gathered from potential users of such a system. These concepts and categories represent advancements on prior theory due to the comprehensive nature of the review. Furthermore, much of the prior literature focuses on case studies. By examining multiple events and cases these concepts are shown to be applicable across a broad variety of events, not individual cases. Due to the space limitation, we will not be able to discuss all the concepts in detail but focus on the four major categories that emerged. For full discussion, please reference the PhD thesis of the first author [24].

Disaster Context-Awareness. Upon speaking with study participants, the importance of situational awareness was immediately emphasized which was reinforced with the literature. Data supported the common understanding that a disaster is made up of dynamic and complex contexts which vary depending on the viewpoint and perspective of many different stakeholders in the regions affected by a disaster. Contextual information provides important details of a situation such as where is the risk, who is at risk, what are the causes of risk, who is available to respond, when did the event occur, what is the rate of environmental change, and much more. The data indicated an understanding of as many key contextual elements of the situation as possible is necessary for responders and decision makers to plan the most effective and efficient response. However, this contextual information must be of high quality to avoid information overload and confusion. Furthermore, a lack of information could lead to problems.

Multi-Agency Response Relationships. Participants and literature sources indicated that not only is the situational information critical to a response, but also the management of relationships amongst responders. Managing relationships is the second category to emerge. Relationships amongst responders are important for establishing leadership roles, leadership structures, communication mechanisms, and identifying roles performed by responding agencies, all of which contribute to the coordination of a disaster response. A history of coordination among agencies will also influence response relationships or the establishment of multi-agency relationships.

Multi-agency Task Management. Thus far the impact of context information and multi-agency relationships have been discussed in relation to disaster response management. Multi-agency task management examines the actual execution of the response. It considers the supporting activities, task management structures, and factors required to organize, plan, and execute the interdependent tasks that make up a disaster response. The main concepts that influence task management include managing the resources, identifying the tasks based on required roles, following pre-established working procedures, maintaining responder safety and readiness, coordination training exercises, and information exchange.

Emergency Support Technology. While technology is not considered a pre-requisite for disaster response, its presence, absence, or fit may affect the performance of the response. The general use of support technology can be affected by many social, technical, and organizational factors regarding its use and perceived effectiveness. These factors include infrastructure dependence, multimedia communication capabilities, systems interoperability, technology costs and funding, systems performance, technology training and expertise, and lastly, responder attitudes towards technology.

3   Conclusion

3.1   Theoretical Contributions

This study uses grounded theory methodology which was uncommonly used in disaster research. By successfully demonstrating the application of grounded theory in conducting research in disaster response areas the method can be used to further develop new theories in the field that are applicable in a more general way. This methodology enabled the consolidation of key concepts from much of the extant case study literature available.

The requirements analysis in this study are beyond traditional IS perspective. Additional issues were examined in the rescue effort which had an overall impact on effective disaster response coordination. For example, the importance of managing multi-agency relationships, and the impact of communication infrastructure damage are issues that would typically not be addressed in an IS paper. However, the identification of these concepts influences overall system design and contributes new requirements whereby information systems can aid in addressing the concerns posed in the non-traditional IS concepts to improve overall emergency response efforts.

The proposed framework also provides a new perspective on decision making. It extends traditional decision support into dynamic situations which require a new way of thinking. Prior literature had expanded decision support contexts from individual decision makers, to include personal, organizational, technical, ethical, and aesthetic perspectives [25]. However, this new framework expands the perspectives to include contextual, multi-party relationship, and task-based coordination aspects into decision making processes. It necessitates the consideration of multiple perspectives in developing the best decisions and corresponding actions for multi-party response.

This study highlights new research directions for decision support and coordination studies. Each perspective and the corresponding factors that contribute to the perspective is an opportunity to extend academic knowledge into new modern decision support areas. Like prior advances in technology expanded opportunities for improved DSS design and implementation, the advancement of ubiquitous mobile communication, environmental sensors, and geographic information systems encourages the growth and distribution of context-aware computing to aid in decision making and coordination as urgent decisions and actions become based on the latest real-time data available.

3.2 Practical Implications

This study can be used to prepare guidelines for the assessment of existing coordination systems, and identify gaps in information systems to be filled enabling better disaster response support. Related to the last point is the identification of how IS can address the issues presented in crisis response coordination with a context-aware multi-party coordination system. An emphasis on improved data mobility and sharing, and improved relationship management functionality are only two of many important requirements to be incorporated in future systems design for more effective crisis response coordination.

The proposed framework can also be applied to many situations beyond emergency response. Many similar operating conditions exist in large scale events and projects, like the Olympic Games, national and international exhibitions [26], and military operations [27]. The requirements and functionality to support the management of such events can benefit from a context-aware multi-party coordination system (CAMPCS) [28] which integrates context with relationships and task-based coordination to support decision making. Although there may not be the same threat of damage and loss of life, these events may have planned schedules to follow and need capabilities to effectively respond to unexpected events triggering a complex chain of reactions.

3.3 Limitations

One of the advantages of this study is the use of mixed data sources with real-world observations and literature sources. However, a major limitation is the type of firsthand data that was available. Firsthand data from an actual large-scale disaster context would be the most desirable. This would have eliminated the need to justify the comparison of smaller emergency data to larger disaster response data. However, the concepts that emerged from the secondhand data reflected important issues that were identified by others, sometimes on a firsthand basis. Furthermore, the common concepts that emerged from both sources provide a future research opportunity. Is there a transitional boundary when an emergency becomes a disaster? When does it occur, what factors influence and determine the transition point, and what changes to information systems are required to support the transition?

Another challenge with this study is the lack of a proof-of-concept system design that can demonstrate the system requirements. This limitation also becomes a future research opportunity. An actual system that could be developed, deployed and tested would further validate the framework and system requirements.

The historical nature of much of the secondhand data left a gap in this research in considering the growing role of social media in disaster response. Some interview participants in the study discussed the growth of social media in their communication activities. Specifically, social media has been used as a method to push information out to the public. While some emergency managers monitored social media in conjunction with other media outlets to track hazard escalations and response needs, the medium was not fully trusted and but represented a growing area and its importance and impact was yet to be determined. This limitation does represent a possible area for future research.

3.4   Future Research

For future research, this study identified several new opportunities based on the identified concepts related to multi-agency coordination. Specifically, opportunities exist to study the representation and management of relationships in real-time decision making structures. This appears to be an area of many opportunities for future research. In addition, the modeling of context in information systems for emergency management applications, and the presentation of such data to emergency responders remains a large opportunity. Many design science proposals ignore key usability assumptions identified in this study. The development of new systems that follow the requirements in this study is an opportunity for new design science studies. Another area of study is the management of dynamic processes in an urgent environment as it relates to project management. Many of the elements of project management implicitly occur when planning disaster response, but the context is so different and dynamic that the tracking mechanisms do not appear to be in place. Improved multi-agency task management systems are another opportunity for future research as they relate to emergency response. Lastly, the applications of social media and mobile computing in emergency management is a new field which is a research opportunity to consider. The ubiquity of mobile computing is creating new opportunities for information gathering and dissemination between the public and responders. The behaviors, motivations, and opportunities for future mobile systems and social media use are growing opportunities for future research.

References

1. Gheytanchi, A., Joseph, L., Gierlach, E., Kimpara, S., Housley, J., Franco, Z.E., Beutler, L. E.: Twelve failures of the hurricane katrina response and how psychology can help. Am. Psychol. **62**, 118–130 (2007)
2. Chen, R., Sharman, R., Rao, H., Upadhyaya, S.: Coordination in emergency response management. Commun. ACM **51**(5), 66–73 (2008)
3. Seeger, M.: Best practices in crisis communication: an expert panel process. J. Appl. Commun. Res. **34**(3), 232–244 (2006)
4. Drabek, T.: Microcomputers and disaster responses. Disasters **15**(2), 186–192 (1991)
5. Mick, S., Wallace, W.: Expert systems as decision aids for disaster management. Disasters **9** (2), 98–101 (1985)
6. Stephenson, R., Anderson, P.: Disasters and the information technology revolution. Disasters **21**(4), 302–334 (1997)
7. Turoff, M., Chumer, M., Van de Walle, B., Yao, X.: The Design of a Dynamic Emergency Response Management Information System (DERMIS). J. Inform. Technol. Theory Appl. **5** (4), 1–36 (2004)
8. Yang, L., Su, G., Yuan, H.: Design principles of integrated information platform for emergency responses: the case of 2008 Beijing Olympic Games. Inform. Syst. Res. **23** (3-part-1), 761–786 (2012)
9. Haddow, G.D., Bullock, J.A.: Introduction to Emergency Management, 2nd edn. Elsevier Inc., Oxford (2006)
10. Petak, W.: Emergency management: A challenge for public administration [Special issue]. Public Adm. Rev. **45**, 3–7 (1985)
11. Salter, J.: Risk management in a disaster management context. J. Contingencies Crisis Manag. **5**(1), 60–65 (1997)
12. Standards Australia/New Zealand. Risk management-Principles and guidelines. (AS/NZS ISO 31000:2009). Sydney, AU & Wellington, NZ: Standards Austrlia/New Zealand (2009)
13. Crondstedt, M.: Prevention, preparedness, response, recovery - an outdated concept? Aust. J. Emerg. Manag. **17**(2), 10–13 (2002)
14. Yuan, Y., Detlor, B.: Intelligent mobile crisis response systems. Commun. ACM **48**(2), 95–98 (2005)
15. Abrahamsson, M., Hassel, H., Tehler, H.: Towards a system-oriented framework for analysing and evaluating emergency response. J. Contingencies Crisis Manag. **18**(1), 14–25 (2010)
16. Van de Walle, B., Turoff, M.: Decision support for emergency situations. Inform. Syst. E-Business Manag. **6**(3), 295–316 (2008)
17. Franco, Z., Zumel, N., Blau, K., Ayhens-Johnson, K., Beutler, L.: Causality, covariates, and consensus in ISCRAM research: towards a more robust study design in a transdisciplinary community. Int. J. Emerg. Manage. **5**(1/2), 100–122 (2008)
18. Urquhart, C., Lehmann, H., Myers, M.: Putting the 'theory' back into grounded theory: guidelines for grounded theory studies in information systems. Inform. Syst. J. **20**(4), 357–381 (2010)
19. Birks, D.F., Fernandez, W., Levina, N., Nasirin, S.: Grounded theory method in information systems research: its nature, diversity and Opportunities. Eur. J. Inform. Syst. **22**, 1–8 (2013)
20. Glaser, B.: Theoretical Sensitivity: Advances in the Methodology of Grounded Theory. Sociology Press, Mill Valley (1978)
21. Glaser, B.: Basics of Grounded Theory Analysis: Emergence vs. Forcing. Sociology Press, Mill Valley (1992)

22. Glaser, B.: Doing Grounded Theory: Issues and Discussions. Sociology Press, Mill Valley (1998)
23. Weber, R., McEntire, D., Robinson, R.: Public/private collaboration in disaster: Implications from the World Trade Center terrorist attacks. Quick Response Research Report 155. Boulder, CO: University of Colorado Natural Hazards Center (2002). http://www.colorado.edu/hazards/research/qr/qr155/qr155.html. Accessed 27 June 2013
24. Way, S.: Requirements analysis for a context-aware multi-agency emergency response system (Doctoral Dissertation) (2013). Retrieved from MacSphere via Open Access Dissertations database: http://hdl.handle.net/11375/13412
25. Mitroff, I., Linstone, H.: Unbounded Mind: Breaking the Chains of Traditional Business Thinking. Oxford University Press, Cary (1993)
26. Meyer, E., Wichmann, D., Büsch, H., Boll, S.: Supporting mobile collaboration in spatially distributed workgroups with digital interactive maps. Mob. Networks Appl. **17**(3), 365–375 (2012)
27. Louverieris, P., Gregoriades, A., Garn, W.: Assessing critical success factors for military decision support. Expert Syst. Appl. **37**(12), 8229–8241 (2010)
28. Way, S., Yuan, Y.: Transitioning from dynamic decision support to context-aware multi-party coordination: a case for emergency response. Group Decis. Negot. (2013). doi:10.1007/s10726-013-9365-3

Convergent Menus of Social Choice Rules

Takahiro Suzuki[1,2(✉)] and Masahide Horita[1]

[1] Department of International Studies, Graduate School of Frontier Sciences,
The University of Tokyo, Kashiwa, Japan
suzuki-t92vh@mlit.go.jp, horita@k.u-tokyo.ac.jp
[2] Construction and Maintenance Management Division, Research Center for
Infrastructure Management, National Institute for Land and Infrastructure
Management, Ministry of Land, Infrastructure,
Transport and Tourism, Tsukuba, Japan

Abstract. Suzuki and Horita [11] proposed the notion of convergence as a new solution for the procedural choice problem. Given a menu of feasible social choice rules (SCRs) F and a set of options X, a preference profile L^0 is said to (weakly) converge to $C \subseteq X$ if every rule to choose the rule (or every rule to choose the rule to choose the rule, and so on) ultimately designates C under a consequential sequence of meta-preference profiles. Although its frequency is shown, for example, under a large society with $F = \{$plurality, Borda, anti-plurality$\}$, a certain failure (trivial deadlock) occurs with small probability. The objective of this article is to find a convergent menu (a menu that can "always" derive the convergence). The results show that (1) several menus of well-known SCRs, such as $\{$Borda, Hare, Black$\}$, are convergent and that (2) the menu $\{$plurality, Borda, anti-plurality$\}$ and a certain class of scoring menus can be expanded so that they become convergent.

Keywords: Convergence · Social choice rules · Procedural choice

1 Introduction

This paper studies the choice of voting rules based on voters' judgments about the procedures. In social choice theory, several concepts have been developed to describe the problem and provide proposed solutions. Many authors have discussed the issue in terms of solution concepts such as self-selectivity (Koray [8]; Koray and Slinko [9]) and self-stability (Barbera and Jackson [2]), among others. These concepts demand that a voting rule choose itself when used as a voting rule to choose the voting rule. Recently, this concepts were extended to menus of voting rules. Houy [6] states that a menu of social choice rules (SCRs) satisfies the condition of first-level stability if, for all preference profiles over the voting rules, the menu includes one and only one SCR that chooses itself. Houy [6] then shows a negative result, saying that no menu of SCRs can satisfy first-level stability and two simple conditions[1]. On the other hand, Diss, Louichi, Merlin, and Smaoui [3] and Diss and Merlin [4] studied the actual probability

---

[1] They are Neutrality, i.e., each SCR in the menu is neutral, and Difference, i.e., there are no pairs of SCRs f, g in the menu that are "identical."

© Springer International Publishing AG 2017
M. Schoop and D.M. Kilgour (Eds.): GDN 2017, LNBIP 293, pp. 47–60, 2017.
DOI: 10.1007/978-3-319-63546-0_4

that a menu of SCRs is stable[2] under the Impartial Culture (IC) and Impartial Anonymous Culture (IAC) models, respectively. Their results show that when the population is (infinitely) large, the probability of stability for the set of {plurality (f_P), Borda (f_B), anti-plurality (f_A)} is 84.49% in the IC model and 84.10% in the IAC model.

As a new solution concept, Suzuki and Horita [11] proposed the notion of *convergence*. In general, a level-1 rule is a SCR for the choice of the set of options X and a level-k rule ($k \geq 2$) is a SCR for the choice of the set of level-($k-1$) rules. For Given a set of feasible SCRs F and a set of options X, a preference profile L^0 is said to (weakly) converge to $C \subseteq X$ if $k \in \mathbb{N}$ exists and every level-k rule ultimately designates C under a consequential sequence of meta-preference profiles. Suzuki and Horita [11] study this notion on a triplet of scoring rules and find that when the population n is large ($n \to \infty$), $F = \{f_P, f_B, f_A\}$, and there are three alternatives, the probability of weak convergence p_{WC} is 98.2% under IC and 98.8% under IAC.

Although p_{WC} is much higher than that of stability for the same menu, a large society still faces a certain type of failure of convergence—trivial deadlock—with small but positive probability: $100\% - 98.2\% = 1.8\%$ (IC) or $100\% - 98.8\% = 1.2\%$ (IAC). The objective of this article is to find ways to avoid such failure. Specifically, we say a menu is *convergent* if a large society with menu F can "always" find convergence (note: a formal definition is given later). Then, formally stated, the research objective is to find non-trivial convergent menus.

This article is organized as follows. Section 2 describes the notation. Section 3 states technical results with some reference to our latest work. The conclusion is stated in Sect. 4. All proofs are in the Appendix.

2   Model

2.1   Basic Definitions

Let $N = \{1, 2, \ldots, n\}$ be a society of n individuals, where $2 \leq n < +\infty$. For any nonempty and finite set A, $\mathcal{L}(A)$ denotes the set of all linear orders over A. A preference profile over A is an n-tuple of linear orders $(L_1, L_2, \ldots, L_n) \in \mathcal{L}(A)^n$, where the ith element L_i is interpreted as individual i's preference.

For any nonempty and finite set of alternatives A, a social choice rule (SCR) f maps the preference profile $L = (L_1, \ldots, L_n) \in \mathcal{L}(A)^n$ into a nonempty subset of A, i.e., $\phi \neq f(L; A) \subseteq A$. A SCR f is called a social choice function (SCF) if it is always singleton-valued. When f is a SCF, with a slight abuse of notation, we write $f(L) = x$ instead of $f(L) = \{x\}$. Let A and B be any nonempty and finite sets with the same cardinalities, $0 < |A| = |B| < \infty$ (A and B can be identical). For any preference profile, $L = (L_1, L_2, \ldots, L_n) \in \mathcal{L}(A)^n$ and a bijection $\sigma : B \to A$, we define a (permuted) preference profile $L^\sigma = (L_1^\sigma, L_2^\sigma, \ldots, L_n^\sigma) \in \mathcal{L}(B)^n$ on B as follows: for all $a, b \in B$ and $i \in N$,

---

$aL_i^\sigma b \Leftrightarrow \sigma(a)L_i\sigma(b)$. We say a SCR is neutral if, for any finite nonempty sets A and B with $|A| = |B|$, alternative $b \in B$, bijection $\sigma : B \to A$, and profile $L \in \mathcal{L}(A)^n$,

$$\sigma(b) \in f(L;A) \Leftrightarrow a \in f(L^\sigma;B).$$

A scoring SCR f for m options is an SCR that assigns to each alternative $s_j (j = 1, 2, \ldots, m)$ points if it is ranked at the jth position in the preferences, where $1 = s_1 \geq s_2 \geq \ldots \geq s_m \geq 0$. Then, $f(L)$ is defined as the set of options with the highest scores. We often express score assignments as $f : [s_1, s_2, \ldots, s_m]$. If $m = 3$, the plurality rule f_P has the assignment $[1, 0, 0]$, the Borda count f_B has the assignment $[1, 1/2, 0]$, and the anti-plurality rule f_A has the assignment $[1, 1, 0]$.

Suppose a society N faces a decision-making problem X, where $2 \leq |X| < \infty$, and it has a menu (the set of feasible SCRs) F, where $2 \leq |F| < \infty$. To make this article self-contained, we give the definition of weak convergence (Definition 5) and some related definitions that are more fully explained in Suzuki and Horita [11]. Then, we define the convergent property (Definition 6).

Definition 1: Level.[3] The level-1 issue is the choice of X using each $f_j \in F$. In this context, each f_j $(j = 1, 2, \ldots, m)$ is called a level-1 SCR and denoted f_j^1 and the level-1 menu is denoted $F^1 = \{f_1^1, \ldots, f_m^1\}$. For any integer $k \geq 2$, the level-k issue is the choice of F^{k-1} using $f_1, f_2, \ldots, f_m$. In this context, each $f_j(j = 1, 2, \ldots, m)$ is called a level-k SCR and denoted f_j^k and the level-k menu is denoted $F^k = \{f_1^k, f_2^k, \ldots, f_m^k\}$.

Definition 2: Class. For any level-1 SCR $f^1 \in F^1$, its class at a level-0 preference profile $L^0 \in \mathcal{L}(X)^n$, denoted $C_{f^1}[L^0]$, is defined as $C_{f^1}[L^0] = f^1(L^0)$.

For any level-$k(\geq 2)$ SCR $f^k \in \mathcal{F}^k$, its class at a level-$0, 1, 2, \ldots, (k-1)$ preference profile $L^0, L^1, \ldots, L^{k-1}$, denoted $C_{f^k}[L^0, L^1, \ldots, L^{k-1}]$, is defined as

$$C_{f^k}[L^0, L^1, \ldots, L^{k-1}] := \bigcup_{g^{k-1} \in f^k(L^{k-1})} C_{g^{k-1}}[L^0, L^1, \ldots, L^{k-2}].$$

Intuitively, the class of $f^k \in \mathcal{F}^k$ represents the ultimate outcome that f^k derives into X. When the sequence $L^0, L^1, \ldots, L^{k-1}$ is obvious in the context, we write it simply as C_{f^k} instead of $C_{f^k}[L^0, L^1, \ldots, L^{k-1}]$.

Definition 3: Preference Extension System. For each $i \in N$, we define $e_i : \mathcal{L}(X) \to \mathcal{L}(\mathfrak{P}(X) \setminus \{\phi\})$ as a preference extension system if it satisfies the following conditions:

(1) For each $a, b \in X$ and $L_i^0 \in \mathcal{L}(X)$, if $(a, b) \in L_i^0$, then $\{a\}e_i(L_i^0)\{b\}$.
(2) For any set $A \subseteq X$ and $b \in X \setminus A$ such that $bL_i^0 a$ for all $a \in A$, $A \cup \{b\}e_i(L_i^0)A$.

---

[3] In this article, we suppose that the society uses the fixed set of SCRs, $f_1, \ldots, f_m$ for any level. The distinction between f_j^1 and f_j^2 by the superscripts is made based on the supposed agenda.

That is, a preference extension system e_i maps each $L_i \in \mathcal{L}(X)$ to a linear order preference over the power set of X (without the empty set). Condition 1 is known in the literature as the Extension Rule (e.g., Barbera, Bossert, and Pattanaik [1]). Almost all the well-known preference extension systems satisfy this condition. Condition 2 says that if the better alternative b is added to A, the new set $A \cup \{b\}$ is evaluated as better than A. This condition is also often referred to in the literature (see, e.g., Gardenfors [5]; Kannai and Peleg [7]). Note that there are many preference extension systems that satisfy these two conditions. Throughout this article we do not specify what kind of e_i each individual has, because the following argument holds regardless.

Definition 4: Consequentially Induced Preference and Profile. For any $i \in N$, $k \in \mathbb{N}$, and $L^0 \in \mathcal{L}(X)^n, L^1 \in \mathcal{L}(F^1)^n, \ldots, L^{k-1} \in \mathcal{L}(F^{k-1})^n$, we define $R_i^k \in \mathcal{W}(F^k)$ as the i's level-k consequentially-induced weak order preference if, for each $f^k, g^k \in F^k$,
$$(f^k, g^k) \in L_i^k \Leftrightarrow (C[f^k : L^0, L^1, \ldots, L^{k-1}], C[g^k : L^0, L^1, \ldots, L^{k-1}]) \in e_i(L_i^0).$$
A linear order $L_i^k \in \mathcal{L}(F^k)$ is called an i's level-k consequentially induced preference (hereafter, level-k CI preference) if it is compatible with the i's level-k consequentially-induced weak order preference. We say $L^0 \in \mathcal{L}(X)^n, L^1 \in \mathcal{L}(F^1)^n, \ldots$, $L^k \in \mathcal{L}(F^k)^n$ is a sequence of CI profiles to level-k if L^j ($j = 1, 2, \ldots, k$) is a CI profile with respect to the previous-level profiles $L^0, L^1, \ldots, L^{j-1}$. We denote by $\mathcal{L}^k[L^0, \ldots, L^{k-1}]$ the set of all level-k CI profiles with respect to a given sequence $L^0, L^1, \ldots, L^{k-1}$ of CI profiles to level $(k-1)$.

When $k = 1$ and F is made up of SCFs only, the CI preference is nothing but the "induced preference" used in the study of self-selective SCRs (Koray [8]). In this sense, the CI preference is a generalization of the induced preference.

Definition 5: Weak Convergence. A level-0 preference profile $L^0 \in \mathcal{L}(X)^n$ is said to weakly converge[4] to $C \subseteq X$ if and only if $k \in \mathbb{N}$ and a sequence of CI profiles to level $(k-1)L^0, L^1, \ldots, L^{k-1}$ exist such that each $f^k \in F^k$ has the same class, i.e., $C[f^k : L^0, L^1, \ldots, L^{k-1}] = C$ for all $f^k \in F^k$.

Throughout this article, we assume the IAC model as the probability model of voting behavior. For a menu F, we say its probability of weak convergence p_{WC} is defined as the probability of occurrence of those level-0 preference profiles that weakly converge.

Definition 6: Convergent Property. We say that F satisfies the asymptotic weak convergent property (we call it simply "convergent property" in this article) if $p_{WC} \rightarrow 1$ as $n \rightarrow \infty$.

---

[4] Whether a profile L^0 weakly converges or not depends on what kind of menu F the society considers, and so it is more precise to say "L^0 weakly converges with respect to the menu F." However, in the subsequent argument, because the menu F is explicit from the context we simply say "L^0 weakly converges to $C \subseteq X$". Also, we do not specify individuals' preference extension systems $\{e_i\}_{i \in N}$ in the definition of convergence. Strictly speaking, a profile L^0 is defined as weakly converging to $C \subseteq X$ if and only if, for combinations of all preference extension systems $\{e_i\}_{i \in N}$, the required sequence of CI profiles exists.

As stated in the Introduction, the objective of this article is to find convergent menus and indeed, several examples of these are shown in the next section. We also note that, based on the definition of the convergent property, the menu $\{f_P, f_B, f_A\}$ is clearly not convergent, because p_{WC} is at most 0.988 as $n \to \infty$.

2.2 An Illustrative Example

Abraham Lincoln (1809–1865), the 16th President of the United States, was elected in 1860. The election is quite interesting from the perspective of social choice theory. There were four candidates running: Abraham Lincoln (L) from Republican Party, John C. Breckinridge (R) from Southern Democratic Party, John Bell (B) from Constitutional Union Party, and Stephen A. Douglas (D) from Northern Democratic Party. Each of them received a significant number of ballots. Although we cannot know for sure the complete preference profile of the citizens at that time, Riker [10] estimates the full preference ranking for each state (Table 1). The table shows the estimated number of people who have the preference lying to its immediate left (for instance, 83000 people are supposed to have the preference of *DLRB*, i.e. Douglas $\succ$ Lincoln $\succ$ Breckinridge $\succ$ Bell).

Table 1. Riker's profile (ballots)

LDRB	0	DLRB	83000	RLDB	0	BLDR	270000
LDBR	450000	DLBR	318000	RLBD	0	BLRD	0
LRDB	0	DRLB	173000	RDLB	104000	BDLR	114000
LRBD	0	DRBL	489000	RDBL	329000	BDRL	28000
LBDR	1414000	DBLR	319000	RBLD	0	BRLD	31000
LBRD	0	DBRL	0	RBDL	413000	BRDL	146000

Based on this table, Riker [10] and Tabarrok and Spector [12] point out that if the citizens' preference profiles had been aggregated using other voting procedures, the result could have been different. We now demonstrate how such discrepancies between voting procedures can be resolved through the notion of weak convergence. Let L_R^0 be Riker's profile over the set of candidates

$$X = \{\text{Lincoln } (L), \text{Douglas } (D), \text{Bell } (B), \text{Breckinridge } (R)\}.$$

Suppose the society has the menu $F = \{f_P, f_B, f_A\}$. It is straightforward to confirm that $f_P^1(L_R^0) = \{L\}$ and $f_B^1(L_R^0) = f_A^1(L_R^0) = \{D\}$. Now, if we take the level-1 CI profile $L^1 = (L_1^1, L_2^1, \ldots, L_n^1)$ as

$$L_i^1 : \begin{cases} f_P^1, f_B^1, f_A^1 & \text{if } LL_i^0 D \\ f_B^1, f_A^1, f_P^1 & \text{if } DL_i^0 L. \end{cases}$$

(it is straightforward to check that this profile L^1 is actually a CI profile), then we have that each level-2 SCR in F^2 chooses $\{f_B^1\}$, which means that L^0 weakly converges to $\{D\}$[5].

Part of the reason for this result is that Douglas wins over Lincoln using the simple majority rule under both profiles. We do not claim that Douglas *should* have been the winner. According to the study of convergence, whether a specific candidate (e.g., Douglas) should be elected depends on what kind of menu the society accepts. For example, if the U.S. citizens at that time thought that f_P was the unique appropriate procedure, i.e., $F = \{f_P\}$, then convergence indicates the victory of Lincoln was appropriate. In the next section, we aim to find a menu F that is convergent and not trivial ($|F| > 1$).

3   Results

Our results are mainly made up of two parts, described in Subsects. 3.1 and 3.2. In Subsect. 3.1, we show that there are several convergent menus made up of well-known SCRs. In Subsect. 3.2, we discuss the possibility of providing the convergent property to a class of non-convergent menus, such as $\{f_P, f_B, f_A\}$.

3.1   Convergent Menus Comprised of Standard Social Choice Rules

Our first result is based on well-known social choice rules. In particular, we consider Hare's procedure f_H, Nanson's procedure f_N, Coomb's procedure f_C, the Maximin procedure f_M, and Black's procedure f_{Bl} in addition to f_P, f_B, and f_A. We denote by $\mathcal{F}$ the set of these eight SCRs, i.e., $\mathcal{F} = \{f_P, f_B, f_A, f_H, f_N, f_C, f_M, f_{Bl}\}$. Taking any three[6] of these eight SCRs to make a menu, e.g., $\{f_H, f_C, f_M\}$, there are ${}_8C_3 = 56$ possible combinations of three different SCRs. Theorem 1 shows that there are 10 convergent menus among these 56 menus.

Lemma 1. Suppose n is sufficiently large. Let $x, y \in X$ and $|F| = 3$ ($F \subseteq \mathcal{F}$). Let $L^0, L^1, \ldots, L^{k-1}$ be a sequence of CI profiles to level $(k-1)$, where $k \in \mathbb{N}$. Suppose

$$\left\{ C_f[L^0, L^1, \ldots, L^{k-1}] \,|\, f \in F^k \right\} = \{\{x\}, \{y\}\}.$$

Then, L^0 weakly converges to x if

$$\#\{i \in N | xL_i^0 y\} > \#\{i \in N | yL_i^0 x\}.$$

---

[5] As a straightforward corollary of Proposition 2 in Suzuki and Horita [11], we can verify that Riker's profile never weakly converges to another subset $C' \subseteq X, C' \neq \{D\}$. So, the victory of Douglas is actually supported in a stronger sense at this profile L^0 under $\{f_P, f_B, f_A\}$.

[6] It is quite hard to investigate menus including four or more SCRs because the Barvinok computer software, which is used to determine the probability introduced below, does not work if there are four or more candidates. This is why we focus only on menus of three SCRs in this subsection, while our next subsection shows a result on a menu of four SCRs.

Theorem 1. Of the 56 menus of SCRs, the following 10 menus of SCRs satisfy the asymptotic weak convergent property, i.e., $p_{WC} \to 1$ as $n \to \infty$,

$$\{f_P, f_N, f_M\}, \{f_A, f_N, f_M\}, \{f_B, f_H, f_{Bl}\}, \{f_B, f_N, f_M\}, \{f_B, f_N, f_{Bl}\}$$
$$\{f_B, f_C, f_{Bl}\}, \{f_B, f_M, f_{Bl}\}, \{f_H, f_N, f_M\}, \{f_N, f_C, f_M\}, \{f_N, f_M, f_{Bl}\}$$

(All proofs are in Appendix A).

Although the existence of convergent menus is not discussed by Suzuki and Horita [11], this theorem shows that they do exist among menus of well-known SCRs. In any of the 10 convergent menus, the probability of occurrence of those profiles that weakly converge is asymptotically one as $n \to \infty$. In other words, a large society equipped with one of these menus can find convergence "without failure," Note that the menus not cited in the theorem do not have the asymptotically weak convergent property. Using the Barvinok computer software implemented by Verdoolaege et al. [13], we can calculate the asymptotic probability of trivial deadlock. The Table 2 shows the calculation results for each of the 56 menus.

Table 2. Probability $(1 - p_{WC})$ for the 56 menus when $|X| = 3$ and $n \to \infty$.

DataNo.	Menu			Prob.		DataNo.	Menu			Prob.	
1	f_P	f_A	f_B	1/84	0.0119	29	f_A	f_H	f_M	11/864	0.0127
2	f_P	f_A	f_H	29/864	0.0336	30	f_A	f_H	f_{Bl}	413/34560	0.0120
3	f_P	f_A	f_N	5/432	0.0116	31	f_A	f_N	f_C	25/1728	0.0145
4	f_P	f_A	f_C	43/1728	0.0249	32	f_A	f_N	f_M	0	0.0000
5	f_P	f_A	f_M	5/432	0.0116	33	f_A	f_N	f_{Bl}	11/4320	0.0025
6	f_P	f_A	f_{Bl}	199/17280	0.0115	34	f_A	f_C	f_M	25/1728	0.0145
7	f_P	f_B	f_H	115/6912	0.0166	35	f_A	f_C	f_{Bl}	113/8640	0.0131
8	f_P	f_B	f_N	1/432	0.0023	36	f_A	f_M	f_{Bl}	11/4320	0.0025
9	f_P	f_B	f_C	1/192	0.0052	37	f_B	f_H	f_N	23/6912	0.0033
10	f_P	f_B	f_M	1/432	0.0023	38	f_B	f_H	f_C	23/3456	0.0067
11	f_P	f_B	f_{Bl}	1/864	0.0012	39	f_B	f_H	f_M	23/6912	0.0033
12	f_P	f_H	f_N	25/1728	0.0145	40	f_B	f_H	f_{Bl}	0	0.0000
13	f_P	f_H	f_C	131/6912	0.0190	41	f_B	f_N	f_C	1/864	0.0012
14	f_P	f_H	f_M	25/1728	0.0145	42	f_B	f_N	f_M	0	0.0000
15	f_P	f_H	f_{Bl}	115/6912	0.0166	43	f_B	f_N	f_{Bl}	0	0.0000
16	f_P	f_N	f_C	7/1728	0.0041	44	f_B	f_C	f_M	1/864	0.0012
17	f_P	f_N	f_M	0	0.0000	45	f_B	f_C	f_{Bl}	0	0.0000
18	f_P	f_N	f_{Bl}	1/864	0.0012	46	f_B	f_M	f_{Bl}	0	0.0000
19	f_P	f_C	f_M	7/1728	0.0041	47	f_H	f_N	f_C	31/6912	0.0045
20	f_P	f_C	f_{Bl}	7/1728	0.0041	48	f_H	f_N	f_M	0	0.0000
21	f_P	f_M	f_{Bl}	1/864	0.0012	49	f_H	f_N	f_{Bl}	23/6912	0.0033
22	f_A	f_B	f_H	241/16128	0.0149	50	f_H	f_C	f_M	31/6912	0.0045
23	f_A	f_B	f_N	67/12096	0.0055	51	f_H	f_C	f_{Bl}	23/3456	0.0067
24	f_A	f_B	f_C	17/1512	0.0112	52	f_H	f_M	f_{Bl}	23/6912	0.0033
25	f_A	f_B	f_M	67/12096	0.0055	53	f_N	f_C	f_M	0	0.0000
26	f_A	f_B	f_{Bl}	181/60480	0.0030	54	f_N	f_C	f_{Bl}	1/864	0.0012
27	f_A	f_H	f_N	11/864	0.0127	55	f_N	f_M	f_{Bl}	0	0.0000
28	f_A	f_H	f_C	185/6912	0.0268	56	f_C	f_M	f_{Bl}	1/864	0.0012

3.2 Convergent expansion

Our next result focuses on the menu $\{f_P, f_B, f_A\}$ again. We already know that this menu is not convergent. It will be shown below that we can make the menu convergent by an expansion.

Let φ be the following SCR:

Condition $(\bigstar)$: $f_P(L) \neq f_B(L)$ and more than half individuals prefer $f_P(L)$ to $f_B(L)$.

$$\varphi(L) := \begin{cases} f_B(L) & \text{if } (\bigstar) \text{ holds} \\ f_P(L) & \text{otherwise.} \end{cases}$$

In words, φ is a SCR which selects $f_P(L)$ or $f_B(L)$ according to the majority rule.

Theorem 2. The menu $\{f_P, f_B, f_A, \varphi\}$ is convergent.

Suppose a large society is equipped with the menu $\{f_P, f_B, f_A\}$. To acquire the convergent property, Theorem 1 demands that the society substitute it with a convergent menu—for instance, $\{f_P, f_N, f_M\}$. On the other hand, Theorem 2 states that the convergent property can also be acquired without abandoning each f_P, f_B, and f_A: the society has only to consider an extra social choice rule, the chair rule φ. Although procedural choice among $\{f_P, f_B, f_A\}$ can sometimes fail to provide the convergence, consideration of an extra procedure makes it go well.

4 Conclusion

In this article, we have discussed the existence of convergent menus, i.e., those menus that "always" result in a large society finding convergence. Although the existence of such menus is not clear in our previous work (Suzuki and Horita [11]) which focuses on scoring rules only, this article shows that there are convergent menus that are made up of well-known social choice rules (Theorem 1). In Subsect. 3.2, we provided an example showing that the expansion of a non-convergent menu $\{f_P, f_B, f_A\}$ with an additional SCR can make it convergent (Theorem 2). It will be an interesting future topic to find whether such expansions are possible for general menu of SCRs.

Acknowledgements. This work was supported by JSPS KAKENHI, grant number 15J07352.

Appendix A

Proof of Lemma 1. Given a menu F and a sequence of CI profiles $L^0, \ldots, L^{k-1}$ which satisfy the stated conditions, let

$$F_x := \{f \in F \mid C_f[L^0, L^1, \ldots, L^{k-1}] = \{x\}\},$$

$$F_y := \{f \in F \mid C[f{:}L^0, L^1, \ldots, L^{k-1}] = \{y\}\},$$

and let $\alpha := |F_x|$ and $\beta := |F_y|$. We label the elements as $F_x = \{g_1, g_2, \ldots, g_\alpha\}$ and $F_y = \{h_1, h_2, \ldots, h_\beta\}$. Also $N_x = \{i \in N \mid x L_i^0 y\}$, $N_y = \{i \in N \mid y L_i^0 x\}$, $n_x = |N_x|$, and $n_y = |N_y|$. Since $\alpha + \beta = 3$, we have two possible cases: (a) $(\alpha, \beta) = (2, 1)$ and (b) $(\alpha, \beta) = (1, 2)$.

(a) The case of $(\alpha, \beta) = (2, 1)$

Define $L^k \in \mathcal{L}^k [L^0, \ldots, L^{k-1}]$ as follows.

$$L_i^k : \begin{cases} g_1, g_2, h_1 & \text{if } i \in N_x \\ h_1, g_1, g_2 & \text{if } i \in N_y. \end{cases}$$

It is easy to see that every $f \in \mathcal{F}$ chooses a subset of $\{g_1, g_2\}$. So, L^0 weakly converges to $\{x\}$.

(b) The case of $(\alpha, \beta) = (1, 2)$

Define $L^k \in \mathcal{L}^k [L^0, L^1, \ldots, L^{k-1}]$ as follows.

$$L_i^k : \begin{cases} g_1, h_1, h_2 & \text{for } (n_y + 1) \text{ individuals in } N_x \\ g_1, h_2, h_1 & \text{for } n_x - (n_y + 1) \text{ individuals in } N_x \\ h_1, h_2, g_1 & \text{for } \lfloor \frac{n}{2} \rfloor - (n_y + 1) \text{ individuals in } N_y \\ h_2, h_1, g_1 & \text{for the other individuals.} \end{cases}$$

Intuitively, L^k is a profile such that the score of g_1 is the largest and the scores of h_1 and h_2 are the smallest. First, it is easy to see that each level-$(k+1)$ $f_P, f_H, f_C, f_{Bl}, f_M$ chooses $\{g_1\}$.

Consider f_B. For simplicity, we denote by $s(f)$ the score of $f \in F^k$ evaluated by f_B. Since each $i \in N_x$ ranks g_1 at the first position and each $i \in N_y$ ranks it at the third, we have $s(g_1) = n_x$. Since $\lfloor \frac{n}{2} \rfloor$ individuals rank h_1 above h_2 and $n - \lfloor \frac{n}{2} \rfloor = \lceil \frac{n}{2} \rceil$ individuals rank h_2 above h_1, we have $s(h_2) \geq s(h_1)$. Furthermore,

$$
\begin{aligned}
s(h_2) &= \frac{1}{2} \left[\{ n_x - (n_y + 1) \} + \left\{ \left\lfloor \frac{n}{2} \right\rfloor - (n_y + 1) \right\} \right] + \left(2n_y + 1 - \left\lfloor \frac{n}{2} \right\rfloor \right) \\
&= \frac{1}{2} n_x + n_y - \frac{1}{2} \cdot \left\lfloor \frac{n}{2} \right\rfloor \\
&\leq \frac{1}{2} n_x + n_y - \frac{1}{2} \cdot n_y \left(\because \left\lfloor \frac{n}{2} \right\rfloor \geq n_y \right) \\
&< n_x = s(g_1) \left(\because n_x > n_y \right).
\end{aligned}
$$

Therefore, $f_B^{k+1}(L^k) = \{g_1\}$. It is straightforward to check that $f_N^{k+1}(L^k) = \{g_1\}$.

Finally, consider f_A. Since n_y individuals rank g_1 at the third position and $(n_y + 1)$ individuals rank h_2 at the third, it follows that $f_A^{k+1}(L^k) \subseteq \{g_1, h_1\}$. Recall that each $f \in \mathcal{F} \setminus \{f_A\}$ chooses $\{g_1\}$ at L^k. If $f_A^{k+1}(L^k) = \{g_1\}$, this implies that L^0 weakly converges to $\{x\}$. If $f_A^{k+1}(L^k) = \{h_1\}$, we can apply the case (a) to the CI sequence $L^0, L^1, \ldots, L^k$ (instead of the sequence $L^0, L^1, \ldots, L^{k-1}$) to find the convergence. Suppose $f_A^{k+1}(L^k) = \{g_1, h_1\}$. Then, it follows that

$$n_x - (n_y + 1) = n_x.$$

This implies that $n_x = 2n_y + 1$. Then, let $M^k \in \mathcal{L}^k[L^0, L^1, \ldots, L^{k-1}]$ as

$$M_i^k : \begin{cases} g_1, h_1, h_2 & \text{for } (n_y + 2)\text{individuals in } N_x \\ g_1, h_2, h_1 & \text{for } (n_y - 1)\text{individuals in } N_x \\ h_1, h_2, g_1 & \text{for } \lfloor \frac{n}{2} \rfloor - (n_y + 2)\text{individuals in } N_y \\ h_2, h_1, g_1 & \text{for the other individuals.} \end{cases}$$

In a similar way, we can check that $f_P, f_H, f_C, f_{Bl}, f_M, f_B, f_N$ chooses $\{g_1\}$ at M^k. Also, we have $f_A^{k+1}(M^k) = \{h_1\}$. So, we can apply the case (a) to the CI sequence $L^0, L^1, \ldots, M^k$ to find the convergence. ∎

Proof of Theorem 1. Let f_1, f_2, f_3 be distinct SCRs among $f_P, f_B, f_A, f_H, f_N, f_C, f_M, f_{Bl}$. When $n \to \infty$ under IAC, it is easy to see that the probability of tied outcomes by some of f_1^1, f_2^1, f_3^1 is negligible. So, we can discuss only $L^0 \in \mathcal{L}(X)^n$ such that each $f_1^1(L^0), f_2^1(L^0), f_3^1(L^0)$ is a singleton.

(1) **The Case of** $\left| \{f_1(L^0), f_2(L^0), f_3(L^0)\} \right| = 2$

Let $\{f_1(L^0), f_2(L^0), f_3(L^0)\} = \{x, y\}$. When $n \to \infty$, the probability of the event

$$\#\{i \in N | x L_i^0 y\} = \#\{i \in N | y L_i^0 x\}$$

is negligible. Hence, we can apply Lemma 1 to derive weak convergence.

(2) **The Case of** $\left| \{f_1(L^0), f_2(L^0), f_3(L^0)\} \right| = 3$

In this case, the level-1 CI profile L^1 is uniquely determined. It is also straightforward to show that the probability of tied outcomes among some of the level-2 SCRs is negligible. If $\left| \{f_1^2(L^1), f_2^2(L^1), f_3^2(L^1)\} \right| = 2$, we can apply Lemma 1 again to derive weak convergence. We next show that $\left| \{f_1^2(L^1), f_2^2(L^1), f_3^2(L^1)\} \right|$ cannot be 3 if the menu F is one of the 56 menus. Suppose to the contrary that it is 3. Note that

$$\mathcal{L}(F^1) = \{f_1 f_2 f_3, f_1 f_3 f_2, f_2 f_1 f_3, f_2 f_3 f_1, f_3 f_1 f_2, f_3 f_2 f_1\}.$$

Let n_j be the number of individuals who have the jth preference. For example, n_1 and n_4 are the numbers of individuals whose level-1 CI preferences are $f_1 f_2 f_3$ and $f_2 f_3 f_1$, respectively.

From now on, the proof is similar for all 10 menus. Let us prove the case of $F = \{f_B, f_H, f_{Bl}\}$. Without loss of generality, we can assume $f_B^2(L^1) = f_1$, $f_H^2(L^1) = f_2$, and $f_{Bl}^2(L^1) = f_3$. With $n_1, \ldots, n_6$, we can rephrase these conditions as follows:

$$f_{Bo}^2(L^1) = f_1^1 : \begin{cases} n_1 + 2n_2 + n_5 > n_3 + 2n_4 + n_6 \\ 2n_1 + n_2 + n_3 > n_4 + n_5 + 2n_6 \end{cases}$$

$$f_H^2(L^1) = f_2^1 : \begin{cases} \begin{cases} n_3 + n_4 > n_1 + n_2 \\ n_5 + n_6 > n_1 + n_2 \\ n_1 + n_3 + n_4 > n_2 + n_5 + n_6 \end{cases} \\ \quad \text{or} \\ \begin{cases} n_1 + n_2 > n_5 + n_6 \\ n_3 + n_4 > n_5 + n_6 \\ n_1 + n_2 + n_5 < n_3 + n_4 + n_6 \end{cases} \end{cases}$$

$$f_{Bl}^2(L^1) = f_3^1 : \begin{cases} \begin{bmatrix} (n_4 + n_5 + n_6 > n_1 + n_2 + n_3 \text{ and } n_2 + n_5 + n_6 > n_1 + n_3 + n_4) \\ \quad \text{or} \\ (n_3 + n_4 + n_6 > n_1 + n_2 + n_5 \text{ or } n_4 + n_5 + n_6 > n_1 + n_2 + n_3) \\ \quad \text{and} \\ (n_1 + n_2 + n_5 > n_3 + n_4 + n_6 \text{ or } n_2 + n_5 + n_6 > n_1 + n_3 + n_4) \\ \quad \text{and} \\ (n_1 + n_2 + n_3 > n_4 + n_5 + n_6 \text{ or } n_1 + n_3 + n_4 > n_2 + n_5 + n_6) \\ \quad \text{and} \\ n_4 + n_5 + 2n_6 > 2n_1 + n_2 + n_3 \\ \quad \text{and} \\ n_2 + 2n_5 + n_6 > n_1 + 2n_3 + n_4 \end{bmatrix} \end{cases}$$

With elementary verification[7], we can see that there is no non-negative integer solution $(n_1, n_2, \ldots, n_6)$ for this system of inequalities.   ∎

Proof of Theorem 2. The probability of tied outcomes among level-1 SCRs can be assumed to be negligible, and so we can expect that each $f_P^1(L^0), f_B^1(L^0), f_A^1(L^0), \varphi^1(L^0)$ is a singleton. If $|F^1(L^0)| \leq 2$, we can apply Lemma 2 of Suzuki and Horita [11] to be assured of weak convergence. Because of the definition of φ, we know that $\varphi(L^0) \subseteq f_P^1(L^0) \cup f_B^1(L^0) \cup f_A^1(L^0)$. So, we can assume $|F^1(L^0)| = 3$. Without loss of generality, we can assume $g_1^1(L^0) = g_2^1(L^0) = \{x_1\}$, $g_3^1(L^0) = \{x_2\}$, and $g_4^1(L^0) = \{x_3\}$, where $F^1 = \{g_1^1, g_2^1, g_3^1, g_4^1\}$.

Let $L^1 \in \mathcal{L}^1[L^0]$ such that all individuals rank g_1^1 above g_2^1. Note that the probability of tied outcomes among some of $f_P^2, f_B^2, f_A^2, \varphi^2$ can also be assumed to be negligible. So, we can expect that $f_P^2(L^1), f_B^2(L^1), f_A^2(L^1), \varphi^2(L^1)$ are also singletons. Then, L^1 is characterized with be the number $n_1, \ldots, n_6$ that expresses the number of individuals who have each specific type of preference, i.e., $g_1^1, g_2^1, g_3^1, g_4^1$ (n_1 individuals), $g_1^1, g_2^1, g_3^1, g_4^1$ (n_2 individuals), $g_3^1, g_1^1, g_2^1, g_4^1$ (n_3 individuals), $g_3^1, g_4^1, g_1^1, g_2^1$ (n_4 individuals), $g_4^1, g_1^1, g_2^1, g_3^1$ (n_5 individuals), and $g_4^1, g_3^1, g_1^1, g_2^1$ (n_6 individuals), where $n = n_1 + n_2 + n_3 + n_4 + n_5 + n_6$. Note also that if $|F^2(L^1)| \leq 2$, then weak convergence is also guaranteed. So, we assume that $|F^2(L^1)| = 3$. At this time, $\varphi^2(L^1)$ is either $f_P^2(L^1)$ or $f_B^2(L^1)$. We can also expect $n_i > 0$ for $i = 1, 2, 3, 4, 5, 6$ when $n \to \infty$, and so we have that $f_A^2(L^1) = \{g_1^1\}$. Now, we have only two possibilities: (1) $f_B^2(L^1) = \varphi^2(L^1) = \{g_3^1\}$, $f_P^2(L^1) = \{g_4^1\}$, and $f_A^2(L^1) = \{g_1^1\}$, or (2) $f_P^2(L^1) = \{g_3^1\}$, $f_B^2(L^1) = \varphi^2(L^1) = \{g_4^1\}$, and $f_A^2(L^1) = \{g_1^1\}$.

---

[7] For actual verification, we used the function "FindInstance" in the software Mathematica.

(1) The case of $f_B^2(L^1) = \varphi^2(L^1) = \{g_3^1\}$, $f_P^2(L^1) = \{g_4^1\}$, and $f_A^2(L^1) = \{g_1^1\}$.

Let $L^2 \in (\mathcal{L}(F^2))^n$ be as follows: $f_A^2, f_B^2, \varphi^2, f_P^2$ (n_1 individuals), $f_A^2, f_P^2, f_B^2, \varphi^2$ (n_2 individuals), $f_B^2, \varphi^2, f_A^2, f_P^2$ (n_3 individuals), $f_B^2, \varphi^2, f_P^2, f_A^2$ (n_4 individuals), $f_P^2, f_A^2, f_B^2, \varphi^2$ (n_5 individuals), and $f_P^2, f_B^2, \varphi^2, f_A^2$ (n_6 individuals). Clearly, we have $L^2 \in \mathcal{L}^2[L^0, L^1]$.

Because $n_1, \ldots, n_6$ are positive, we obtain $f_A^3(L^2) = \{f_B^2\}$. Furthermore, $f_P^2(L^1) = \{g_4^1\}$ and so the plurality score of g_4^1 is greater than those of g_1^1 and g_3^1:

$$\begin{cases} n_5 + n_6 > n_1 + n_2. \\ n_5 + n_6 > n_3 + n_4. \end{cases}$$

This also shows that the plurality score of f_P^2 is greater than those of f_A^2 and f_B^2 at L^2. Hence, we have that $f_P^3(L^2) = \{f_P^2\}$. Next, we show that $f_B^3(L^2) = \{f_B^2\}$.
Because $f_B^2(L^1) = \{g_3^1\}$, the Borda scores at L^1 are as follows:

$$s_B(g_3^1) > s_B(g_1^1) \Leftrightarrow n_3 + n_4 + \frac{2}{3}n_6 + \frac{1}{3}n_1 > n_1 + n_2 + \frac{2}{3}(n_3 + n_5) + \frac{1}{3}(n_4 + n_6).$$

$$s_B(g_3^1) > s_B(g_4^1) \Leftrightarrow n_3 + n_4 + \frac{2}{3}n_6 + \frac{1}{3}n_1 > n_5 + n_6 + \frac{2}{3}n_4 + \frac{1}{3}n_2.$$

At L^2, we have:

$$s_B(f_P^2) = n_5 + n_6 + \frac{2}{3}n_2 + \frac{1}{3}n_4.$$

$$s_B(f_B^2) = n_3 + n_4 + \frac{2}{3}(n_1 + n_6) + \frac{1}{3}(n_2 + n_5).$$

$$s_B(f_A^2) = n_1 + n_2 + \frac{2}{3}n_5 + \frac{1}{3}n_3.$$

$$s_B(\varphi^2) < s_B(f_B^2).$$

These equations show that $s_B(f_B^2) > \max\{s_B(f_P^2), s_B(f_A^2), s_B(\varphi^2)\}$.

(2) The case of $f_B^2(L^1) = \{g_3^1\}$, $f_P^2(L^1) = \varphi(L^1) = \{g_4^1\}$, and $f_A^2(L^1) = \{g_1^1\}$. $f_B^2(L^1) = \{g_3^1\}$, and so the score of g_3^1 at L^1 is strictly greater than those of g_1^1 and g_4^1. Formally, we have that:

$$\begin{aligned} n_3 + n_4 + \frac{2}{3}n_6 + \frac{1}{3}n_1 &> n_1 + n_2 + \frac{2}{3}(n_3 + n_5) + \frac{1}{3}(n_4 + n_6). \\ n_3 + n_4 + \frac{2}{3}n_6 + \frac{1}{3}n_1 &> n_5 + n_6 + \frac{2}{3}n_4 + \frac{1}{3}n_2. \end{aligned} \tag{1}$$

Let $L^2 \in \mathcal{L}^2[L^0, L^1]$ be such that:

$$n_1 \text{ individuals:} f_A^2, f_B^2, f_P^2, \varphi^2.$$
$$n_2 \text{ individuals:} f_A^2, f_P^2, \varphi^2, f_B^2.$$
$$n_3 \text{ individuals:} f_B^2, f_A^2, f_P^2, \varphi^2.$$
$$n_4 \text{ individuals:} f_B^2, f_P^2, \varphi^2, f_A^2.$$
$$n_5 \text{ individuals:} f_P^2, \varphi^2, f_A^2, f_B^2.$$
$$n_6 \text{ individuals:} f_P^2, \varphi^2, f_B^2, f_A^2.$$

In words, this is a consequentially induced preference where everyone ranks f_P^2 above φ^2. Similar to (1), we can check that $f_P^3(L^2) = f_A^3(L^2) = \{f_P^2\}$. Furthermore, the scores of $f_A^2, f_P^2, f_B^2, \varphi^2$ evaluated by f_B^3 are as follows:

$$s_B(f_A^2) = n_1 + n_2 + \frac{2}{3}n_3 + \frac{1}{3}n_5.$$

$$s_B(f_B^2) = n_3 + n_4 + \frac{2}{3}n_1 + \frac{1}{3}n_6.$$

$$s_B(f_P^2) = n_5 + n_6 + \frac{2}{3}(n_2 + n_4) + \frac{1}{3}(n_1 + n_3).$$

With the inequalities (1) we have that $s_B(f_B^2) > s_B(f_A^2)$. Note that ties between f_B^2, f_P^2 occur only if

$$n_3 + n_4 + \frac{2}{3}n_1 + \frac{1}{3}n_6 = n_5 + n_6 + \frac{2}{3}(n_2 + n_4) + \frac{1}{3}(n_1 + n_3).$$

This event is negligible as $n \to \infty$.   ∎

References

1. Barbera, S., Bossert, W., Pattanaik, P.K.: Ranking Sets of Objects. Kluwer Academic Publishers, Dordrecht (2004)
2. Barbera, S., Jackson, M.O.: Choosing how to choose: self-stable majority rules and constitutions. Q. J. Econ. **119**, 1011–1048 (2004)
3. Diss, M., Louichi, A., Merlin, V., Smaoui, H.: An example of probability computations under the IAC assumption: the stability. Math. Soc. Sci. **64**, 57–66 (2012). doi:10.1016/j.mathsocsci.2011.12.005
4. Diss, M., Merlin, V.: On the stability of a triplet of scoring rules. Theor. Decis. **69**, 289–316 (2010). doi:10.1007/s11238-009-9187-6
5. Gardenfors, P.: Manipulation of social choice functions. J. Econ. Theor. **13**, 217–228 (1976)
6. Houy, N. (2004). A note on the impossibility of a set of constitutions stable at different levels∗

7. Kannai, Y., Peleg, B.: A note on the extension of an order on a set to the power set. J. Econ. Theor. **32**, 172–175 (1984)
8. Koray, S.: Self-selective social choice functions verify arrow and Gibbard-Satterthwaite theorems. Econometrica **68**, 981–995 (2000)
9. Koray, S., Slinko, A.: Self-selective social choice functions. Soc. Choice Welf. **31**, 129–149 (2006)
10. Riker, W.H.: Liberalism Against Populism: A Confrontation Between the Theory of Democracy and the Theory of Social Choice. Freeman, San Francisco (1982)
11. Suzuki, T., Horita, M.: Plurality, Borda count, or anti-plurality: regress convergence phenomenon in the procedural choice. In: Bajwa, D., Koeszegi, S., Vetschera, R. (eds.) Group Decision and Negotiation. Theory, Empirical Evidence, and Application. LNBIP, vol. 274, pp. 43–56. Springer, Cham (2017). doi:10.1007/978-3-319-52624-9_4
12. Tabarrok, A., Spector, L.: Would the Borda count have avoided the Civil War? J. Theor. Polit. **11**, 261–288 (1999). doi:10.1177/0951692899011002006
13. Verdoolaege, S., Beyls, K., Bruynooghe, M., et al.: Analytical computation of ehrhart polynomials and its applications for embedded systems. In: 2nd Work Optimizations DSP Embedded System, pp. 1–12 (2004)

Fuzzy Group Decision-Making
for the Remediation of Uranium Mill Tailings

Antonio Jiménez-Martín[1], Hugo Salas[1], Danyl Pérez-Sánchez[2],
and Alfonso Mateos[1]($^{(\boxtimes)}$)

[1] Decision Analysis and Statistics Group,
Universidad Politécnica de Madrid, Boadilla del Monte, Spain
{ajimenez,amateos}@fi.upm.es, hugo.salas.pablos@alumnos.upm.es
[2] Departamento de Medio Ambiente, CIEMAT, Madrid, Spain
d.perez@ciemat.es

Abstract. In this paper, we propose an approach for evaluating remediation alternatives at the Zapadnoe uranium mill tailings site in a group decision-making context. The approach relies on both interval values and a linguistic term scale to valuate the alternative impacts and ordinal information about the relative importance of criteria. Monte Carlo simulation techniques are used to exploit imprecision to compute a fuzzy dominance matrix for each DM, taking into account the corresponding ordinal information about weights. Then a *fuzzy dominance measuring method* is used to derive the corresponding rankings of remediation alternatives. Finally, they are aggregated taking into account their relative importance to reach a consensus ranking.

1 Introduction

Nearly sixty years of uranium mining and milling has resulted in a variety of smaller and larger environmental liabilities in several of the member states of the European Union (EU) and the former Soviet Union (USSR) [1]. As a consequence of growing awareness on the part of the national authorities over the last two or three decades and fueled by increased public concern, the closure of processing activities led to the necessity to remediate tailings deposits in order to reduce their impact on health and the environment to acceptable and sustainable levels.

The selection of a preferred remediation alternative is a complex decision-making problem in which other factors aside from the radiological and chemical toxicity impacts of the wastes have to be taken into account. A theoretical framework based on multi-criteria decision analysis (MCDA) was proposed in [2–4] to analyze remediation alternatives for the Zapadnoe uranium mill tailings, in which a fuzzy linguistic term scale was proposed to evaluate remediation alternatives and to quantify preferences.

In this paper, we consider the same scenario, the safety assessment of Zapadnoe uranium mill tailings remediation, and extend the theoretical model in [2–4] to solve the problem. First, although vagueness/uncertatinty about the remediation alternative performances can be represented again using a fuzzy linguistic term scale, uniformly distributed performance intervals can be also used.

© Springer International Publishing AG 2017
M. Schoop and D.M. Kilgour (Eds.): GDN 2017, LNBIP 293, pp. 61–69, 2017.
DOI: 10.1007/978-3-319-63546-0_5

Regarding imprecision concerning DM preferences, imprecise component utility functions can be built for attributes whose performance intervals are available. Additionally, ordinal information can be provided by each DM regarding weights, i.e., each DM provides an attribute importance ranking. Finally, the theoretical model in [2–4] has also been extended to a group decision-making context, involving researchers from the International Atomic Energy Agency IAEA, the NRPA (Norway), the IGC (Ukraine) and the CIEMAT (Spain).

We propose using Monte Carlo simulation techniques to exploit imprecision in the decision-making parameters to compute a fuzzy dominance matrix for each DM taking into account the corresponding ordinal information about weights. We then use a *fuzzy dominance measuring method* [5] to derive the corresponding rankings of remediation alternatives, which are, finally, aggregated taking into account their relative importance to reach a consensus ranking.

The paper is structured as follows. In Sect. 2 we introduce the problem scenario and structure. The assessment of imprecise DM preferences is introduced in Sect. 2.1. The process for deriving a consensus ranking on the basis of the computation of the individual fuzzy dominance matrices by means of Monte Carlo simulations is explained in Sect. 3. Finally, some conclusions are given in Sect. 4.

2   Problem Scenario and Structure

The Zapadnoe uranium mill tailings site (Zapadnoe tailings) is situated in the south-western part of the main industrial site of the former Pridneprovsky Chemical Plant, located at Dneprodzerzhinsk (Ukraine). The tailings site operated from 1949 until 1954. The total volume of waste was $3.5 \times 10^5 m^3$, and the total activity was 1.8×10^{14} Bq [6]. There are two aquifers at the Zapadnoe tailings site. The groundwater flow in the alluvial aquifer is directed to the north towards the Konoplyanka and the Dnepr rivers.

The context for a safety assessment of the Zapadnoe tailings site was described in [6]. The safety assessment itself was carried out by *ecomonitor* and *geo-eco-consulting* following the steps set out in the ENSURE II project, coordinated by the *Swedish radiation safety authority* (SSM) with the aimed of providing assistance to Ukraine in the remediation of uranium contaminated territories and facilities at the Dnieprodzerchynsk industrial site.

The operational history of the tailings site, its engineering features, as well as the chemical, physical and radioactive characteristics of the waste materials in the tailings are reported in [6].

To identify the criteria to be incorporated into the analysis, experts and stakeholders taking part in ENSURE II project were consulted and the literature on applications of MCDA for the remediation of uranium mill tailing sites was reviewed (see e.g. [7–9]). An objective hierarchy applicable to remediation options for Zapadnoe tailings was built, see Fig. 1. There are three main criteria at the highest level: radiological impact, social impact and economic impact.

The *radiological impact* is split into three subobjectives. *Public radiological impact* refers to the doses received by the population due to external exposure, inhalation and ingestion. It differentiates the doses received by the population during the implementation of the remediation alternative and after the implementation, leading to two new subobjectives, respectively.

To measure the public radiological impact after the implementation, we take the current dose (11.04 mSv/year) as a reference point and identify the reduction that the implementation of the remediation alternatives would involve. The attribute range is then (0, 11.04). Public radiological impact during the implementation accounts for the percentage increment over the current dose.

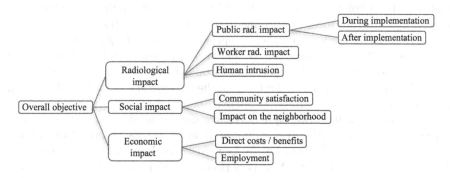

Fig. 1. Objective hierarchy.

Radiological impact on workers refers to radiation doses received by workers as a consequence of the process of implementing a remediation alternative. It accounts for the external dose (radiation exposure on the surface of the tailings), the doses received by inhalation and by ingestion. We consider the number of hours it would take to implement the respective remediation alternative, the number of workers and the radiation doses received per hour. The fuzzy linguistic scale is used to quantify this objective.

Finally, *Human intrusion* refers to the radiation received by intruders at the Zapadnoe tailings site, and it again accounts for the external dose, the doses received by both inhalation and ingestion. To measure this objective, we take the current dose (1.964 mSv/h) as a reference point and identify the percentage reduction that the implementation of the remediation alternatives would involve.

Social impact is split into community satisfaction and the impact on neighborhoods or regions. *Community satisfaction* refers to how a remediation alternative is perceived by individuals belonging to a critical group living in the area, and the *impact on the neighborhood* accounts for the impact on the local community as a whole, including dust, light, noise, odor and vibration during the remediation works and associated with traffic, including operations during day, night and weekend shifts. The fuzzy linguistic scale is used to quantify both social objectives.

Under *Economic impact, Direct costs/benefits* refer to the costs of implementing and maintaining a remediation alternative (manpower, consumables...). It also accounts for the possible direct economic benefits (e.g., sale of waste materials for reutilization). *Employment* refers to job creation through the implementation of a remediation alternative and afterwards. Short- and long-term jobs are taken into account.

2.1   Remediation Alternatives and Valuation

There are three possible strategies. *No action* refers to the natural evolution of the situation without intervention. *Treating or stabilizing the tailing sin situ* is applicable to radiological, non-radiological and mixed contamination. One of the most straightforward means of dealing with contaminated sites appears to be to isolate them from human and other receptors by constructing physical barriers. *Removing the tailings*. Contaminated materials are removed from the site and transferred to a designated disposal site. Conditioning may be required before disposal. Generally, any method relying on the removal of contaminated material is likely to involve substituting the removed material with clean soil.

As mentioned before, the vagueness surrounding the alternative impacts can be represented using uniformly distributed impact intervals or terms from a fuzzy linguistic term scale (VL: Very low, L: Low, M-L: Medium-Low, M: Medium, M-H: Medium-High, H: High and VH: Very High), in which each term has associated a trapezoidal fuzzy number, see Fig. 2.

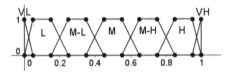

Fig. 2. Fuzzy linguistic term scale.

Table 1 shows the impacts associated with the remediation alternatives. For attribute X_3, it takes 24 and 30 months to recover and remove the tailings, respectively, whereas the radiation doses received are 2.3375 and 2.7115 mSv/year and the number of workers is 62 in both cases. Consequently, we used the fuzzy impacts shown in Table 1. For attribute X_7, the costs associated with recovery and removal are similar since the higher costs associated with removal are cancelled out by the sale of waste materials for reutilization. Regarding X_8, two people are currently responsible for the security of the tailings site. As mentioned before for attribute X_3, 62 employees would be required over a 24- and 30-month period, respectively, for recovery and removal. Security guards would not be necessary in either of the above two cases.

Imprecise component utility functions were built for attributes whose associated impacts were provided by means of value intervals. The function shown in Fig. 3 corresponds to *Public radiation impact (after implementation)*. It is a decreasing utility function since the best value corresponds to no radiation

Table 1. Alternative remediation impacts

	A_1: No action	A_2: New covering	A_3: Removing
X_1: Public rad. imp. (during)	0%	$25 \pm 5\%$	$45 \pm 5\%$
X_2: Public rad. imp. (after)	11.04	(4.416, 5.52)	(0, 0.202)
X_3: Worker rad. impact	No impact	M	M-H
X_4: Human intrusion	0%	$92.5 \pm 2.5\%$	100%
X_5: Community satisfaction	L	VH	H
X_6: Impact on the neighborhood	No impact	VL	M-L
X_7: Direct costs/benefits	No impact	M-H	H
X_8: Employment	M-L	M-H	H

and the worst one is 11.24 mSv/year, which is the current value for such criteria before implementing any remediation alternative. Note that a trapezoidal fuzzy number is derived from the imprecise impact of the *New covering* alternative, (4.416, 5.52), accounting for the class of utility functions. The resulting trapezoidal fuzzy number is (0.5, 0.6, 0.65, 0.713).

Linear component utility functions decreasing and increasing in [0,100] were considered for X_1: Public rad. imp. (during) and X_4: Human intrusion, respectively.

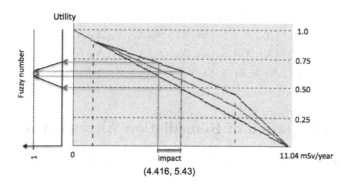

Fig. 3. Fuzzy number construction.

Table 2 shows the fuzzy valuations associated with the remediation alternatives.

We took into account the preferences of four experts, who participated in problem solving, belonging to the Unit Waste & Environmental Safety Section of International Atomic Energy Agency (IAEA), the Emergency Preparedness & Environmental Radioactivity Norwegian Radiation Protection Authority (NRPA), the Institute of Geological Sciences (Ukraine) and the Energy,

Table 2. Fuzzy preferences regarding alternative impacts

	A_1: No action	A_2: New covering	A_3: Removing
X_1	(1,1,1,1)	(0.7,0.7,0.8,0.8)	(0.5,0.5,0.6,0.6)
X_2	(0,0,0,0)	(0.5,0.6,0.65,0.71)	(0.8,0.84,1,1)
X_3	(1,1,1,1)	(0.35,0.45,0.55,0.65)	(0.15,0.25,0.35,0.45)
X_4	(0,0,0,0)	(0.9,0.9,0.95,0.95)	(1,1,1,1)
X_5	(0,0.05,0.15,0.25)	(0.95,1,1,1)	(0.75,0.850.95,1)
X_6	(1,1,1,1)	(0.95,1,1,1)	(0.55,0.65,0.75,0.85)
X_7:	(1,1,1,1)	(0.15,0.25,0.35,0.45)	(0,0.05,0.15,0.25)
X_8	(0.15,0.25,0.35,0.45)	(0.55,0.65,0.75,0.85)	(0.75,0.85,0.95,1)

Environment and Technology Research Centre (CIEMAT, Spain), respectively. The four DMs were considered equally important.

Besides, ordinal information about weights was also provided by the DMs, i.e., an attribute importance ranking, arranged in descending order from the most to the least important attribute, see Table 3.

Table 3. Ordinal information concerning weights

DM	ordinal information
DM_1	$\{X_1, X_2, X_3, X_4\} > \{X_5, X_6\} > \{X_7, X_8\}$
DM_2	$\{X_5, X_6\} > \{X_7, X_8\} > \{X_1, X_2, X_3, X_4\}$
DM_3	$\{X_7, X_8\} > \{X_5, X_6\} > \{X_1, X_2, X_3, X_4\}$
DM_4	$X_2 > X_1 > X_3 > X_4 > X_6 > X_5 > X_8 > X_7$

3   Fuzzy Evaluation of Remediation Alternatives

The remediation alternatives can be evaluated by means of an additive model,

$$u(A_i) = \sum_{j=1}^{8} w_j v_{ij},$$

where w_j represents the relative importance of the attribute X_j and v_{ij} is the valuation of the alternative in the attribute X_j. Note that $\sum_j w_j = 1$ and $w_j \geq 0$.

However, we have considered fuzzy valuation for remediation alternatives and ordinal information from each DM concerning weights. A *fuzzy dominance measuring method* (FDMM) [5] is then used to take advantage of the above imprecise information to derive a ranking of the remediation alternatives. To do this, a *fuzzy dominance matrix* must be first computed for each DM:

$$\widetilde{D}^l = \begin{pmatrix} - & \widetilde{D}^l_{12} & \widetilde{D}^l_{13} \\ \widetilde{D}^l_{21} & - & \widetilde{D}^l_{23} \\ \widetilde{D}^l_{31} & \widetilde{D}^l_{32} & - \end{pmatrix},$$

where superscript l refers to the l-th DM and

$$\widetilde{D}^l_{ks} = min\{\widetilde{v^l(A_k)} - \widetilde{v^l(A_s)}\} = min\{ \oplus_{j=1}^{8} w^l_j \otimes (\widetilde{v}^l_{kj} \ominus \widetilde{v}^l_{sj}) \}$$
$$s.t.$$

$$w^l_{(1)} \geq w^l_{(2)} \geq ... \geq w^l_{(8)}, \ \textstyle\sum_j w^l_j = 1$$

where the fuzzy arithmetic proposed in [10] is used.

Regarding the *min* operator in the objective function, we propose an approach based on a similarity function [11], which compares the fuzzy numbers under consideration based on their similarity to $(-1, -1, -1, -1)$. Note that crisp pairwise dominance values are within $[-1, 1]$.

Monte Carlo simulation techniques can be used to compute the above fuzzy pairwise dominance values. Specifically, the pairwise dominance between alternatives A_k and A_s for the l-th DM, can computed as follows:

1. Do $\widetilde{D}^{lMIN}_{ks} = (1, 1, 1, 1)$.
2. Randomly generate a weight vector accounting for the ordinal information provided by the l-th DM, see Table 3. To do this, we select seven independent random numbers from a uniform distribution on $(0, 1)$, and rank these numbers. Suppose the ranked numbers are $1 \geq r_7 \geq ... \geq r_2 \geq r_1 > 0$. The differences between consecutive ranked numbers are then used as the target weights $w^l_8 = 1 - r_7, w^l_7 = r_7 - r_6, ..., w^l_1 = r_1$. The resulting weights will sum 1 and be uniformly distributed in the weight space.
3. Compute $\widetilde{D}^l_{ks} = \oplus_{j=1}^{8} w^l_j \otimes (\widetilde{v}^l_{kj} \ominus \widetilde{v}^l_{sj})$. If $\widetilde{D}^l_{ks} < \widetilde{D}^{lMIN}_{ks}$ on the basis of their similarity to $(-1, -1, -1, -1)$, then $\widetilde{D}^{lMIN}_{ks} = \widetilde{D}^l_{ks}$. If $stop - criterion$, return $\widetilde{D}^{lMIN}_{ks}$. Otherwise, go to Step 2.

For instance, the fuzzy dominance matrix for the first DM $(\widetilde{D}^1)$ is:

$$\begin{pmatrix} - & (-0.35, -0.32, -0.274, -0.224) & (-0.316, -0.274, -0.181, -0.124) \\ (0.049, 0.074, 0.098, 0.116) & - & (-0.149, -0.125, -0.035, 0.0004) \\ (-0.023, 0.001, 0.026, 0.045) & (-0.226, -0.2, -0.148, -0.098) & - \end{pmatrix}.$$

After computing the fuzzy dominance matrix, $\widetilde{D}^l$, we apply the FDMM [5] to derive a remediation alternative ranking as follows: First, we compute the strength of dominance of alternative A_k by adding the trapezoidal fuzzy numbers in the kth row of $\widetilde{D}^l$, denoted by $\widetilde{d}^l_k$.

Next, we compute a *dominance intensity*, DI^l_k, for each alternative A_k as the proportion of the positive part of the fuzzy number $\widetilde{d}^l_k$ by the distance of the fuzzy number to zero.

Table 4. Dominance intensities and remediation alternative rankings for DMs.

	DM_1	DM_2	DM_3	DM_4	Consensus
1st	$A_2(-0.005)$	$A_2(-0.021)$	$A_1(-0.366)$	$A_3(-0.276)$	$A_2(-0.218)$
2nd	$A_3(-0.162)$	$A_3(-0.427)$	$A_2(-0.443)$	$A_2(-0.404)$	$A_3(-0.344)$
3rd	$A_1(-0.526)$	$A_1(-0.775)$	$A_3(-0.510)$	$A_1(-1.228)$	$A_1(-0.724)$

Specifically, if all of $\widetilde{d}_k^l$ is located to the left of zero, then DI_k^l is minus the distance of $\widetilde{d}_k^l$ to zero, $-D(\widetilde{d}_k^l, 0, f)$, because there is no positive part in $\widetilde{d}_k^l$. If all of $\widetilde{d}_k^l$ is located to the right of zero, then $DI_k^l = D(\widetilde{d}_k^l, 0, f)$, because there is no negative part in $\widetilde{d}_k^l$. Finally, if $\widetilde{d}_k^l$ includes the zero in its base, then the fuzzy number will have a part on the right of zero $(\widetilde{d}_k^{lR})$ and another part on the left of zero $(\widetilde{d}_k^{lL})$. DI_k^l is the proportion that represents $\widetilde{d}_k^{lR}$ with respect to $\widetilde{d}_k^l$ by $D(\widetilde{d}_k^l, 0, f)$ less the proportion that represents $\widetilde{d}_k^{lR}$ with respect to $\widetilde{d}_k^l$ by $D(\widetilde{d}_k^l, 0, f)$, i.e.,

$$DI_k = (\widetilde{d}_k^{lR} - \widetilde{d}_k^{lL}) \times D(\widetilde{d}_k^l, 0, f).$$

$D(\widetilde{d}_k^l, 0, f)$ refers to Tran and Duckstein's [12] distance adapted for the distance of a trapezoidal fuzzy number $\widetilde{d}_k$ to a constant (specifically 0). Thanks to the presence of function f the DM attitude toward risk is incorporated to the analysis.

The alternatives are ranked accordingly to the dominance intensities, DI_k^l. Table 4 shows the dominance intensities and the resulting rankings for the DMs under consideration. Finally, the *order explicit algorithm* [13] is used to derived a consensus ranking (last column in Table 4). A_2: *New covering* is first ranked in the consensus ranking, followed by A_3: *Removing* and A_1: *No action*.

4   Conclusions

We have proposed an approach to evaluate remediation alternatives in the Zapadnoe uranium mill tailings site in a group decision-making context. Thanks to the possibility of using a linguistic term scale and interval values (with the corresponding imprecise component utility functions) to valuate the impacts associated with the alternatives in conjunction with ordinal information about the relative importance of criteria, DMs are comfortable with the proposed approach. Their individual rankings are then aggregated to output a consensus ranking.

Acknowledgment. The paper was supported by the Spanish Ministry of Economy and Competitiveness MTM2014-56949-C3-2-R.

References

1. Vrijen, J., Hahne, R., Barthel, R., Tunger, B., Deissmann, G.: Situation concerning Uranium mine and mill tailings in an Enlarged EU. Final report, EU Contract TREN/04/ NUCL/07.39881 (2006)
2. Jiménez-Martín, A., Martín, M., Pérez-Sánchez, D., Mateos, A., Dvorzhak, A.: A MCDA framework for the remediation of Zapadnoe uranium mill tailings: a fuzzy approach. J. Mult-Valued Log. S. Comput. **26**, 529–540 (2016)
3. Pérez-Sánchez, D., Jiménez, A., Mateos, A., Dvorzhak, A.: A fuzzy MCDA framework for remediation of a Uranium tailing. In: Intelligent Systems and Decision Making for Risk Analysis and Crisis Response: Proceeding of 4th International Conference on Risk Analysis and Crisis Response, pp. 81–88 (2013)
4. Pérez-Sánchez, D., Jiménez, A., Mateos, A., Dvorzhak, A.: Fuzzy MCDA for remediation of a Uranium tailing. In: Merkel, B.J., Arab, A. (eds.) Uranium - Past and Future Challenges, pp. 113–122. Springer, Cham (2015). doi:10.1007/ 978-3-319-11059-2_13
5. Jiménez, A., Mateos, A., Sabio, P.: Dominance intensity measure within fuzzy weight oriented MAUT: an application. Omega **41**, 397–405 (2013)
6. Bugay, D., Voitsekhovitch, O., Lavrova, A., Skalskyy, A., Tkachenko, E.: Definition of the assessment context and scenarios for the selected cases. Data analyses and data collection. Technical report, Project Ensure-II: WP 2. Ecomonitor (2012)
7. Goldammer, W., Nüsser, A., Bütow, E., Lühr, H.-P.: Integrated assessment of radiological and non-radiological risks at contaminated sites. Mathematische Geologie **3**, 53–72 (1999)
8. Bonano, E.J., Apostolakis, G.E., Salter, P.F., Ghassemi, A., Jennings, S.: Application of risk analysis and decision analysis to the evaluation, ranking and selection of environmental remediation alternatives. J. Hazard. Mater. **7**, 35–57 (2000)
9. Kunze, C., Walter, U., Wagner, F., Schmidt, P., Barnekow, U., Gruber, A.: Environmental impact and remediation of Uranium tailing and waste rock dumps at Mailuu Suu (Kyrgyzstan). In: International Conference on Mine Closure and Environmental Remediation, Wismut (2007)
10. Banerjee, S., Kumar Roy, T.: Arithmetic operations on generalized trapezoidal fuzzy number and its applications. Turkish J. Fuzzy Syst. **3**, 16–44 (2012)
11. Vicente-Cestero, E., Jiménez-Martín, A., Mateos, A.: Similarity functions for generalized trapezoidal fuzzy numbers: an improved comparative analysis. J. Intell. Fuzzy Syst. **28**, 821–833 (2015)
12. Tran, L., Duckstein, L.: Comparison of fuzzy numbers using a fuzzy number measure. Fuzzy Sets Syst. **130**, 331–341 (2002)
13. Lin, S., Ding, J.: Integration of ranked lists via cross entropy Monte Carlo with applications to mRNA and microRNA studies. Biometrics **65**, 9–18 (2009)

Conflict Resolution

University of Hohenheim

Stream Introduction: Conflict Resolution

Liping Fang[1,2], Keith W. Hipel[2,3,4], and D. Marc Kilgour[2,5]

[1] Department of Mechanical and Industrial Engineering, Ryerson University,
350 Victoria Street, Toronto, ON M5B 2K3, Canada
lfang@ryerson.ca
[2] Department of Systems Design Engineering, University of Waterloo,
Waterloo, ON N2L 3G1, Canada
kwhipel@uwaterloo.ca, mkilgour@wlu.ca
[3] Centre for International Governance Innovation, Waterloo, ON N2L 6C2,
Canada
[4] Balsillie School of International Affairs, Waterloo, ON N2L 6C2, Canada
[5] Department of Mathematics, Wilfred Laurier University, Waterloo, ON N2L 3C5,
Canada

1 Overview

Strategic conflicts, ranging from pure competition to highly collaborative situations, arise whenever humans interact, individually or in groups. The design of new methodologies that can assist analysts in understanding strategic conflicts and provide strategic support to negotiators has benefited many decision makers. Novel theoretical issues are being addressed in conjunction with the construction of flexible software systems for implementation of decision technologies that can be utilized by both researchers and practitioners. The resulting techniques and associated decision support systems have been used to study strategic conflicts in diverse areas including energy projects, environmental management, global warming, the food crisis, economic disparities, international trade and regional wars. The major objective of the Stream on Conflict Resolution is to provide a forum for the development of formal approaches to conflict resolution and/or insightful applications in a range of fields.

With respect to GDN 2017, researchers working on important topics connected to conflict resolution have contributed one full paper for this volume in which the authors tackle the complex South China Sea dispute using attitude analysis.

As is evident, the theory and application of conflict resolution methodologies in group decision and negotiation can address a wide range of interesting problems. Many more research successes are to be expected in the future.

Evolutional Analysis for the South China Sea Dispute Based on the Two-Stage Attitude of Philippines

Peng Xu, Haiyan Xu$^{(\boxtimes)}$, and Shawei He

College of Economics and Management, Nanjing University of Aeronautics
and Astronautics, Nanjing, Jiangsu, China
2270717678@qq.com, {xuhaiyan,shaweihe}@nuaa.edu.cn

Abstract. Due to different attitudes of ex-president and president of Philippines for the South China Sea dispute, the different equilibria of this conflict arose to facilitate the negotiation between China and Philippines. The evolutional conflict models resulted from decision makers' attitude based on option prioritization under the graph model for conflict resolution are constructed and analyzed in this paper. Compared with the first stage of the South China Sea dispute, the equilibrium of the second stage conflict is different from the first one because current presidential attitude of Philippines is not negative for Chinese government. The two-stage equilibria provide the valuable information that helps decision makers to choose suitable attitude that can be better to understand and resolve the conflict.

Keywords: Evolution analysis · Attitude · Option prioritization · Graph model for conflict resolution · The South China Sea dispute

1 Introduction

China is the first country to discover and name the Spratly Islands, and continues to exercise sovereignty over the Spratly Islands. However, after World War II, Southeast Asian countries illegally occupied the Spratly Islands, such as the Philippines, Vietnam, Malaysia and so on. Recently, due to the South China Sea arbitration case [1], the situation between the Philippines and China in the South China Sea is becoming increasingly tense, which results in the conflict between the Philippines and China. However, the South China Sea dispute between China and the Philippines undergone a great turning, due to the attitude of the current president of the Philippines towards China with positive manner. Therefore, it is important to study the evolution of conflict equilibrium caused by the change of decision makers (DMs)' attitude. In this paper, the two-stage evolutional models within the framework of the graph model for conflict resolution (GMCR) [2] including the DMs' attitude are presented by studying the South China Sea dispute based on the two-stage attitude of the Philippines.

GMCR was proposed by Kilgour et al. [2] to solve strategic conflict in 1987, whose greatest advantage is the need for very little conflict information compared with classical game theory. GMCR is a formal analysis method for conflict developed on the

© Springer International Publishing AG 2017
M. Schoop and D.M. Kilgour (Eds.): GDN 2017, LNBIP 293, pp. 73–85, 2017.
DOI: 10.1007/978-3-319-63546-0_6

basis of the classical game theory [3] and the metagame theory [4], which has a strict mathematical structure and was applied in many fields [5–7]. Then, in order to depict a variety of behaviors for DMs in a conflict, a series of stability definitions were presented, including Nash stability [8, 9], General Metarationality (GMR) [4], Symmetric Metarationality (SMR) [4] and Sequential stability (SEQ) [10]. However, above definitions have not considered the attitude of DMs.

In 1993, Fang et al. [11] first suggested that the attitude of DMs should be introduced into the conflict, because the DMs' attitude will affect the DMs' preferences [12] and the result of conflict due to DMs' preferences generated by the subjective judgment of DMs. In the previous studies, the preference of DMs was determined only by their own interests. When the attitude is taken into account, the preference of DMs may be changed, because the DMs' preference is generated not only by his own interests but also by the opponent's interests. Accordingly, in 2007, Inohara concluded the three types of attitudes [13], and defined four basic attitude stability of RNash, RGMR, RSEQ, RSMR [14, 15]. However, the correlative attitude definitions proposed by Inohara are based on logical representations, whose process for calculating attitude stability is more complex. Subsequently, Sean B transformed the logical definition of attitude stability into matrix expression [16–18], which provides a great convenience for us to calculate the attitude stabilities and laid the foundation for future system development.

However, above attitude definitions proposed by Inohara and Sean B are based on the preference of states, but the preference of states is difficult to obtain in the complex conflict. So Xu et al. proposed the preference of attitude based on option prioritization [19] in 2016, because the quantity of states is more than the quantity of options in a complex conflict and option prioritization is easy to get for a user. Let the numbers of states and options be m and k, respectively. Then, m and k satisfy the equation $m = 2^k$ that shows the preference of attitude based on option prioritization is very convenient to generate. But, the above conflicts are assumed with fixed attitudes that cannot describe the evolutional conflict due to DMs' changed attitude. Therefore, existing attitude theory is unable to accurately analyze and predict the conflict. In this paper, the evolutional graph model within the framework of the attitude theory based on option prioritization is presented by studying the South China Sea dispute in view of the two-stage attitude of the Philippines, which helps DMs to better understand and resolve the conflict.

2 Attitude Based on Option Prioritization Under GMCR

2.1 The Graph Model for Conflict Resolution

A graph model for conflict resolution is a 4-tuple $(N, S, (A_i)_{i \in N}, (>i, \sim i)_{i \in N})$, where N: the set of DMs ($|N| \geq 2$), S: the set of all states in the conflict ($|S| \geq 2$), (S, A_i): DM i's graph (S: the set of all vertices, $A_i \in S \times S$: the set of all arcs such that $(s, s) \notin A_i$ for all $s \in S$ and all $i \in N$), and $(>_i, \sim_i)$: DM i's preferences on S.

Briefly, the conflict includes four elements (DMs, States, Moves, Preference). DMs denote all participants in the conflict; States indicate the strategy combination of all

DMs; Moves represent DM can move between two states by changing his own strategy; Preference denotes the rank of states according to DMs' preference. Here, the states and DMs' moves are represented by the vertices and directed arcs in the graph model, respectively. Therefore, a graph constitutes a natural construct in which to model a conflict.

With respect to preferences, for $s, t \in S$, $s >_i t$ means that DM i prefers state s to t, while $s \sim_i t$ indicates that DM i is indifferent between s and t.

2.2   Attitude Based on Option Prioritization

Definition 1 (Attitudes): Attitude is a stable psychological tendency of an individual to a particular object (person, idea, emotion, or event).

This psychological tendency contains the subjective evaluation and the behavioral tendencies of the individual. In a conflict, the DMs' preferences are generated by the subjective evaluation of DMs, hence the DMs' attitude should be taken into account when we calculate DMs' preference.

Inohara divides the attitude of DMs into three kinds (Positive, Negative and indifferent) [13], in which, the positive, indifferent and negative attitude of DM i towards DM j are denoted by $e_{ij} = +$, $e_{ij} = 0$ and $e_{ij} = -$, respectively.

The following related definitions are defined based on option prioritization [18, 20–22], which is a method to get preference of states. In the option prioritization, the DM i's option statements are denoted by $L_i (i = 1, 2, 3, \ldots n)$. Under the L_i, the preference of DM i can be obtained, denoted by $P_i (i = 1, 2, 3 \ldots n)$.

Definition 2 (Positive attitude option statements): If $e_{ij} = +$, DM i will make an option statements that is beneficial to DM j, denoted by $L_i(e_{ij} = +) = L_j$.

Here, the option statements of DM i under positive attitude towards DM j are the same as the DM j's option statements, which is beneficial for DM j.

Definition 3 (Negative attitude option statements): If $e_{ij} = -$, DM i will make a option statements that is harmful to DM j, denoted by $L_i(e_{ij} = -) = -L_j$.

Under the negative attitude towards DM j, the option statements of DM i are the same as the opposite of DM j's option statements, which is injurious for DM j.

Definition 4 (Indifferent attitude option statements): If $e_{ij} = 0$, DM i doesn't care his option statements in this attitude, denoted by $L_i(e_{ij} = 0) = I$.

Definition 5 (Attitude option statements):

$$L_{ij} = \begin{cases} L_j & \text{if } e_{ij} = + \\ -L_j & \text{if } e_{ij} = - \\ I & \text{if } e_{ij} = 0 \end{cases}$$

Here, L_{ij} denotes DM i's option statements at corresponding attitude.

Definition 6 (Attitude preference): According to L_{ij}, the attitude preference of DM i is obtained, denoted by T_{ij}. For s, $t \in S$ and $i \in N$, $t \in T_{ij}(s)$ if and only if $t >_i s$ satisfies T_{ij}.

Definition 7 (Total attitude preference): For s, $t \in S$ and $i \in N$, $t \in T_i^+(s)$ if and only if $t \in T_{ij}(s)$ for all $j \in N$, then we call total attitude preference.

Here, DM i's total attitude preference satisfies all attitude preferences. In other words, the state in the intersection of all DM i's attitude preferences is what DM i want to reach (total attitude preference).

Definition 8 (Set of less or equally preferred states at total attitude): For s, $t \in S$, and $i \in N$, $t \in T_i^{-=}(s)$ if and only if $t \notin T_i^+(s)$.

Definition 9 (Reachable list): For $i \in N$, $s \in S$, DM i's reachable list from state s is the set $\{t \in S | (s, t) \in A_i\}$, denoted by $R_i(s) \subset S$.

The reachable list is a record of all the states that a given DM can reach from a specified starting state in one step.

Definition 10 (Unilateral improvement list for a DM at attitude): For s, $t \in S$ and $i \in N$, $t \in T_i^*(s)$ if and only if $t \in R_i(s)$ and $t \in T_i^+(s)$.

According the definition, the state in $T_i^*(s)$ is reachable and preferable for DM i at initial state s.

2.3   Stability Concepts of Attitude

Definition 11 (Relational Nash stability—RNash): If $T_i^*(s) = \varnothing$, then $s \in S_i^{RNash}$.

A state s is RNASH stable for DM i iff i has no unilateral improvement moves at attitude from state s, namely, DM i doesn't want to move or cannot reach to the preferred states from state s.

Definition 12 (Relational general metarationality—RGMR): If for all $h \in T_i^*(s)$, and $R_j(h) \cap T_i^{-=}(s) \neq \varnothing$, then $s \in S_i^{RGMR}$.

DM i will not move to the unilateral improvement state at attitude if i finds that the opponent j could make a move regardless of the benefit to himself that sanctions i's moves.

Definition 13 (Relational symmetric metarationality—RSMR): If for all $h \in T_i^*(s)$, exist $y \in R_j(h) \cap T_i^{-=}(s)$ and $z \in T_i^{-=}(s)$ for all $z \in R_i(y)$, then $s \in S_i^{RSMR}$.

If DM i cannot escape the sanction on i's unilateral improvement moves at attitude by DM j, then DM i likes to stay on initial state s. RSMR presumes one step more foresight than RGMR, because it evaluates not only the response by the opponent to DM i's moves but also the counterresponse from DM i.

Definition 14 (Relational sequential stability—RSEQ): If for all $h \in T_i^*(s)$, and $T_j^*(h) \cap T_i^{-=}(s) \neq \varnothing$, then $s \in S_i^{RSEQ}$.

Here, DM i's all potential unilateral improvement moves at attitude are sanctioned by DM j's unilateral improvement moves at attitude. Hence, RSEQ is the same as RGMR except that DM i takes the benefit of his own into consideration at time in sanction.

3 The Evolutional Two-Stage Dispute Between China and the Philippines for the South China Sea

The dispute between China and the Philippines in the South China Sea mainly revolves around the ownership of the islands sovereignty and maritime demarcation issues. Dispute began in the early 1950s, the U.S. military forces in the Philippines Subic Bay turned Huangyan Island into a place for military exercise without authorization, ignored China's sovereignty. [23] On April 30, 1997, the Philippine two representatives boarded the Huangyan Island and set the Philippine national flag on it. On April 10, 2012, the Philippine navy Palawan captured Chinese fishermen in Huangyan Island waters. [24] On March 26, 2013, the Philippines unilaterally submitted the South China Sea dispute to the International Tribunal for the Law. On July 12, 2016, Arbitration tribunal made an illegal and invalid arbitration. China repeatedly states that the Aquino III of the Republic of Philippines unilaterally filed arbitration is no jurisdiction, and China will not accept and recognize [25].

But with the change of the Philippine president, the South China Sea dispute has undergone a major turning. The new president, Rodrigo Duterte, said he was more willing to promote reconciliation with China, rather than international arbitration. Furthermore, the new president does not agree with the pro-American policy of former president completely, who will consider more to promote reconciliation with China, restoring investment, trade, tourism and other aspects with China. Because the Philippines found that USA just want to utilize him to implement the Asia-Pacific rebalancing strategy, and did not provide any substantive assistance to him [26].

3.1 Basic Modeling

Decision Makers, Options and Feasible States. In this conflict, there are two DMs: China (C), Philippines (P). China has two options: 1. Settle territorial disputes through negotiation (Negotiation); 2. Resolve the dispute in the South China Sea through military power (Declaration of war). The Philippines also has two options: 3. Return the islands to China and jointly develop the rich resources in South China Sea (Return the islands); 4. Seek help from other countries or organizations, and jointly confront China, such as USA, Vietnam, Malaysia and so on (Ask for help). (Presented in the left of Table 1).

Logically, there are $2^4 = 16$ of states because the number of options is 4. But there are some states are not reasonable. For example, if China chooses option 1, option 2 cannot be selected, and the Philippines cannot choose option 3 and option 4 at same time. Lastly, we will get 9 feasible states after removing unreasonable states. (Shown in the right of Table 1).

Graph Model of Conflict. In Fig. 1, there are China's moves and the Philippine moves depicting the movement that DMs unilaterally control between two states, and the dot indicates 9 feasible states, the directed arc denotes DM can transfer between the two states by changing his own strategy. The reachable list of DM is naturally produced from graph model.

Table 1. Feasible states of the dispute between China and Philippines

DMs	Options	Feasible states								
C	1. Negotiation	N	N	N	N	N	N	Y	Y	Y
	2. Declaration of war	N	N	N	Y	Y	Y	N	N	N
P	3. Return the islands	N	N	Y	N	N	Y	N	N	Y
	4. Ask for help	N	Y	N	N	Y	N	N	Y	N
Label		S1	S2	S3	S4	S5	S6	S7	S8	S9

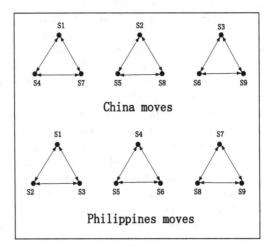

Fig. 1. Graph model of the dispute between China and Philippines

Option Statements. For China, China most likes the Philippines returns the islands, and China hopes to solve this conflict through negotiation, not war. Because if China fights with the Philippines, USA will have an excuse to intervene this dispute, and the Philippines will fight with China by seeking help from other international organizations or counties. Therefore, China's option statements from most preferred to least is 3, 1, −2, −4. Here, 3 denotes China likes option 3, and −2 denotes China likes the opposite of option 2.

For the Philippines, the Philippines does not want China to solve this dispute through military power, because the Philippine military power is far less than China's and USA has not given clear attitude to help the Philippines in military. But the Philippines also doesn't like to return the Spratly Islands to China, because the Spratly Islands have a wealth of resources and the important military strategy position. Hence the Philippines' option statements from the most preferred to least preferred is −2, −3, 4, −1.

3.2 Attitude Modeling and Analysis for the First Stage Dispute

Attitude. For China, his position on the South China Sea issue is to uphold the sovereignty of the South China Sea islands and the surrounding waters, who hope to resolve this issue through negotiation according international law. The main purpose of China is to defend the integrity of national sovereignty, and China insists on the principle, putting aside disputes and developing together, for the rich resources in the South China Sea before solving this dispute. China just considers his own benefit and is not evil intention to the Philippines, in other words, China has a positive attitude for himself and an indifferent attitude towards the Philippines [27].

For the Philippines, he does not want to return the island, because the South China Sea has a wealth of resources, and is the only way which must be passed for the East Asian countries shipping trade. Moreover, the Philippines was a former US colony, is the alliance relationship with the United States during World War II. As China's military power in the South China Sea gradually increase, the Philippines hopes to join forces with USA and other counties to suppress China's power in the South China Sea. Thus, the Philippines has a positive attitude for himself and a negative attitude towards China. [28] (Presented in the Table 2)

Table 2. Attitudes of the first stage dispute between China and Philippines

	China	Philippines
China	+	0
Philippines	−	+

Attitude Option Statements. According to definition 5 and the attitude among DMs, the attitude option statements of DMs are obtained presented in the Table 3. For example, $L_{CC} = L_C(e_{CC} = +)$ denotes China's attitude option statements under the attitude ($e_{CC} = +$) is same to China's option statements.

Attitude Preference. The attitude preference of DM is generated by corresponding attitude option statements presented in the Table 4. For example, T_{CC} denotes China's attitude preference under the attitude ($e_{CC} = +$).

Table 3. Attitude option statements of the first stage dispute between China and Philippines

DMs	Attitude option statements (L_{ij})	
China	$L_{CC} = L_C\ (e_{CC} = +)$	$L_{CP} = I(e_{CP} = 0)$
	3	Doesn't care
	1	
	−2	
	−4	
Philippines	$L_{PC} = -L_C(e_{PC} = -)$	$L_{PP} = L_P(e_{PP} = +)$
	−3	−2
	−1	−3
	2	4
	4	−1

Table 4. Attitude preference of the first stage dispute between China and Philippines

DMs	Attitude		Attitude preference (T_{ij})
China	$e_{CC} = +$	T_{CC}	S9 > S3 > S6 > S7 > S8 > S1 > S2 > S4 > S5
	$e_{CP} = 0$	T_{CP}	Doesn't care
Philippines	$e_{PC} = -$	T_{PC}	S5 > S4 > S2 > S1 > S8 > S7 > S6 > S3 > S9
	$e_{PP} = +$	T_{PP}	S2 > S8 > S1 > S7 > S3 > S9 > S5 > S4 > S6

Total Attitude Preference. The total attitude preference should satisfy every attitude preference (Presented in the Table 5). For example, $T_C^+ = T_{CC} \cap T_{CP}$ denotes China's total attitude preference satisfies attitude preferences T_{CC} and T_{CP}.

Attitude Stability Analysis. Based on above total attitude preference, graph model and the definition of attitude stabilities, the equilibrium of this conflict is calculated and displayed in the Table 6, in which the "$\sqrt{}$" denotes the state is stable for DM under the corresponding stability, and "Eq" means an equilibrium that is a stable state for all DMs. It is clear that S8 is an equilibrium under four kinds of attitude stabilities, that is to say S8, China wants to solve this dispute through negotiation and the Philippines chooses to join other countries and organizations to boycott China, is the possible result of the first stage dispute.

Table 5. Total attitude preference of the first stage dispute between China and Philippines

States	$T_C^+ = T_{CC} \cap T_{CP}$	$T_P^+ = T_{PC} \cap T_{PP}$
S1	S3, S6, S7, S8, S9	S2
S2	S1, S3, S6, S7, S8, S9	Null
S3	S9	S1, S2, S7, S8
S4	S1, S2, S3, S6, S7, S8, S9	S5
S5	S1, S2, S3, S4, S6, S7, S8, S9	Null
S6	S3, S9	S1, S2, S4, S5, S7, S8
S7	S3, S6, S9	S1, S2, S8
S8	S3, S6, S7, S9	S2
S9	Null	S1, S2, S3, S7, S8

Table 6. Attitude stability of the first stage dispute between China and Philippines

States	RNASH			RGMR			RSMR			RSEQ		
	C	P	Eq	C	P	Eq	C	P	Eq	C	P	Eq
S1					√		√				√	
S2		√			√			√			√	
S3				√	√	√	√	√	√	√		
S4					√		√				√	
S5		√			√			√			√	
S6				√			√			√		
S7	√			√	√	√	√	√	√	√		
S8	√	√	√	√	√	√	√	√	√	√	√	√
S9	√			√	√	√	√	√	√	√		

3.3 Attitude Modeling and Analysis for the Second Stage Dispute

With the Philippine presidential transition, the South China Sea dispute has undergone a major turning due to the changed attitude of the Philippines (Fig. 2). The Philippine attitude towards China may transform from the negative to indifferent or positive, because the new President Rodrigo Duterte advocates peaceful means to solve this dispute, he publicly declared willing to jointly develop the rich resources in the South China Sea with China, emphasized his willingness to cooperate with China in economic cooperation and his reluctance to confront China. Another important reason is that the Philippines finds USA did not provide any substantive (Military, Political) assistance to him. Accordingly, the Philippines is not willing to continue to confront China with USA, and the result of this dispute may be changed [26].

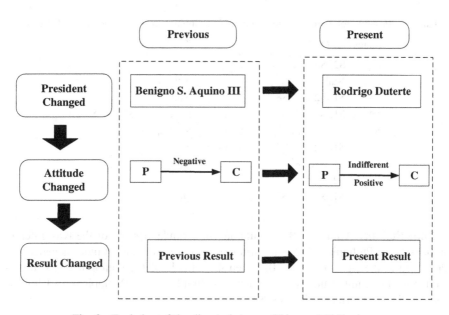

Fig. 2. Evolution of the dispute between China and Philippines

Table 7. Possible attitude of Philippines in the second stage of dispute

	China	Philippines
① Philippines	0	+
② Philippines	+	+

Possible Attitude of the Philippines. There are two kinds attitude for the Philippines in the second stage dispute, the first is that the Philippine attitude to China transforms from negative to indifferent, the second is that the Philippine attitude to China transforms from negative to positive. (Presented in the Table 7).

Possible Attitude Preference and Total Attitude Preference of the Philippines.
Two possible attitude preferences and total attitude preferences are presented in the
Tables 8 and 9, respectively.

Table 8. Possible attitude preferences of Philippines in the second stage of dispute

Possible attitude preference for Philippines (T_{ij})	
① $T_{PC}(e_{PC} = 0)$	Doesn't care
$T_{PP}(e_{PP} = +)$	S2 > S8 > S1 > S7 > S3 > S9 > S5 > S4 > S6
② $T_{PC}(e_{PC} = +)$	S9 > S3 > S6 > S7 > S8 > S1 > S2 > S4 > S5
$T_{PP}(e_{PP} = +)$	S2 > S8 > S1 > S7 > S3 > S9 > S5 > S4 > S6

Table 9. Possible total attitude preferences of Philippines in the second stage of dispute

States	① (T_P^+)	② (T_p^+)
S1	S2, S8	S8
S2	Null	Null
S3	S1, S2, S7, S8	Null
S4	S1, S2, S3, S5, S7, S8, S9	S1, S2, S3, S7, S8, S9
S5	S1, S2, S3, S7, S8, S9	S1, S2, S3, S7, S8, S9
S6	S1, S2, S3, S4, S5, S7, S8, S9	S3, S9
S7	S1, S2, S8	Null
S8	S2	Null
S9	S1, S2, S3, S7, S8	Null

Attitude Stability Analysis. By calculating the attitude stability of the two different
attitudes of the Philippines, the equilibrium under corresponding attitudes is generated.
(Presented in the right of Table 10) When the attitude of the Philippines to China is
indifferent, the equilibrium is S8. When the attitude of the Philippines to China is
positive, the equilibrium is S7, S8, S9.

3.4   Results Evolution Analysis for the South China Sea Dispute

As shown in the Table 10, when the attitude of the Philippines towards China changes
from negative to indifferent, the equilibrium of the conflict does not change, is still S8,
which indicates the Philippines will continue to seek assistance from other international
organizations or counties. Because when the Philippine attitude towards China becomes
indifferent, out of the consideration of his own interests, such as developing the rich
resources and weakening China's control in the South China Sea, the Philippines will
continue to maintain the previous strategy, so the equilibrium is still S8.

When the attitude of the Philippines towards China changes from negative to
positive, the equilibrium of conflict is S7, S8 and S9. S7 indicates that the Philippines
has given up seeking allies to confront China. S9 denotes the Philippines chooses to

Table 10. The result evolution process for the South China Sea dispute

DMs	Options	Previous equilibrium	Present equilibrium		
		$e_{PP} = +$ $e_{CP} = -$	$e_{PP} = +$ $e_{CP} = 0$	$e_{PP} = +$ $e_{CP} = +$	
		S8	S8	S8	S7	S9
C	1. Negotiation	Y	Y	Y	Y	Y
	2. Declaration of war	N	N	N	N	N
P	3. Return the islands	N	N	N	N	Y
	4. Ask for help	Y	Y	Y	N	N

return the island, and jointly develops the rich resources in the South China Sea with China. These three equilibria are likely to be the final solution. But we find that if the equilibrium changes from S8 to S7 and then to S9, the Philippine strategy is gradually beneficial to China. The Philippines firstly gives up to seek allies to confront China, and then chooses to return the island, which reflects some inner links among those equilibria. Perhaps we can further determine the final resolution based on some other properties of DMs' attitude, which will be researched in future.

4 Conclusion and Future Work

In this paper, the two-stage evolutional conflict including the Philippines' different attitudes based on GMCR is presented using the two-stage evolution of the South China Sea dispute. The results show that when the attitude of the Philippines towards China changes from negative to indifferent, out of the consideration of his own interests, the Philippines will maintain the previous strategy, the equilibrium does not change. When the attitude of the Philippines towards China transforms from negative to positive, the equilibria is beneficial to China. Accordingly, China should adopt some peaceful approaches to solve this dispute, including laying disputes aside and developing toge-ther, promoting strategic mutual trust and interest integration, strengthening industry collaboration with the Philippines and so on, which contributes to the transformation of Philippine attitude towards China from negative to positive. The analysis of this evo-lutional dispute not only provides valuable information for DMs, but also helps them to better understand and resolve conflicts, and promotes DMs cooperating well.

But this evolutional analysis for the South China Sea dispute is based on the changed attitude of the Philippines only. In fact, there may be other important factors affecting the South China Sea dispute evolution. For example, the USA may be involved in this dispute with the US new president's attitude, which may result in the South China Sea dispute to evolve again. Considering these factors into the evolutional analysis for the South China Sea dispute will be very valuable.

Acknowledgements.   The authors appreciate financial support from the National Natural Science Foundation of China (Nos. 71071076, 71471087, and 61673209).

References

1. Jinming, L.I., University, X.: The South China Sea Arbitration Case: A Public Opinion War Waged by U.S. & Philippines. Pacific Journal (2016)
2. Kilgour, D.M., Hipel, K.W., Fang, L.: The graph model for conflicts. Automatica **23**(1), 41–55 (1987)
3. Von Neumann, J., Morgenstern, O.: Theory of Games and Economic Behavior. Princeton University Press, Princeton (1944)
4. Howard, N.: Paradoxes of Rationality: Theory of Metagames and Political Behavior. MIT press, Cambridge (1971)
5. Ma, J., Hipel, K.W., De, M.: Strategic analysis of the James Bay Hydroelectric Dispute in Canada. Can. J. Civ. Eng. **32**(5), 868–880 (2005)
6. Hipel, K.W., Kilgour, D.M.: Comparison of the analytic network process and the graph model for conflict resolution. J. Syst. Sci. Syst. Eng. **14**(3), 308–325 (2005)
7. Sheikhmohammady, M., Hipel, K.W., Asilahijani, H., Kilgour, D.M.: Strategic Analysis of the Conflict over Iran's Nuclear Program, pp. 1911–1916 (2009)
8. Nash, J.F.: Equilibrium points in n-person games. Proc. Nat. Acad. Sci. **36**(1), 48–49 (1971)
9. Nash, J.F.: Noncooperative games. Ann. Math. **54**, 286–295 (1951)
10. Fraser, N.M., Hipel, K.W.: Solving complex conflicts. IEEE Trans. Syst. Man Cybern. **9**(12), 805–817 (1979)
11. Fang, L., Hipel, K.W., Kilgour, D.M.: Interactive Decision Making: The Graph Model for Conflict Resolution. Wiley, New York (1993)
12. Walker, S., Hipel, K.W., Inohara, T.: Attitudes and preferences: Approaches to representing decision maker desires. Appl. Math. Comput. **218**(12), 6637–6647 (2012)
13. Inohara, T., Hipel, K.W., Walker, S.: Conflict analysis approaches for investigating attitudes and misperceptions in the war of 1812. J. Syst. Sci. Syst. Eng. **16**(2), 181–201 (2007)
14. Inohara, T., Yousefi, S., Hipel, K.W.: Propositions on interrelationships among attitude based stability concepts. In: IEEE International Conference on Systems, Man and Cybernetics, pp. 2502–2507. IEEE (2008)
15. Inohara, T., Hipel, K.W.: Interrelationships among attitude based and conventional stability concepts within the graph model for conflict resolution. In: Proceedings of the IEEE SMC, pp. 1130–1135 (2009)
16. Walker, S.B., Hipel, K.W., Xu, H.: A Matrix representation of attitudes in conflicts. IEEE Trans. Syst. Man Cybern. Syst. **43**(6), 1328–1342 (2013)
17. Walker, S.B.: Conflicting Attitudes in Environmental Management and Brownfield Redevelopment. Ipb. (2012)
18. Walker, S.B., Hipel, K.W., Inohara, T.: Strategic decision making for improved environmental security: coalitions and attitudes. J. Syst. Sci. Syst. Eng. **18**(4), 461–476 (2009)
19. Xu, H., Xu, P., Sharafat, A.: Matrix Representation of Attitude based on Option Prioritization for Two-DM Conflicts. Submitted to Transactions of Nanjing University of Aeronautics & Astronautics for review
20. Hipel, K.W., Kilgour, D.M., Fang, L., Peng, X.: The decision support system GMCR in environmental conflict management. Appl. Math. Comput. **83**(2–3), 117–152 (1997)
21. Hou, Y., Jiang, Y., Xu, H.: Option prioritization for three-level preference in the graph model for conflict resolution. In: Kamiński, B., Kersten, Gregory E., Szapiro, T. (eds.) GDN 2015. LNBIP, vol. 218, pp. 269–280. Springer, Cham (2015). doi:10.1007/978-3-319-19515-5_21
22. Yu, J., Hipel, K.W., Kilgour, D.M., Zhao, M.: Option prioritization for unknown preference. J. Syst. Sci. Syst. Eng. **25**(1), 39–61 (2016)

23. Liao, Q.: A review of studies on disputes over the South China Sea in recent years. J. Wuhan Commercial Serv. Coll. **30**(3), 83–87 (2016). (In Chinese)
24. Xinhuanet. http://news.xinhuanet.com/english/china/2012-04/18/c_131533604_2.htm
25. BBC NEWS. http://www.bbc.com/news/world-asia-26781682
26. Phoenix New Media. http://news.ifeng.com/a/20160520/48810179_0.shtml
27. Phoenix New Media. http://news.ifeng.com/a/20160713/49348667_0.shtml
28. Wei, H.: The impact of Asia Pacific rebalancing strategy for the South China Sea dispute. Theory Res. **19**, 38–39 (2014). (In Chinese)

Emotions in Group Decision and Negotiation

University of Hohenheim

Stream Introduction: Emotions in Group Decision and Negotiation

Bilyana Martinovski

Department of Computer and System Sciences, University of Stockholm,
Stockholm, Sweden
bilyana@dsv.su.se

1 Overview

Empirical observations show that emotions play an important role in decision taking even if they are restricted by rigid rules or settings. In a group context, emotions are not only subjective but also inter-subjective experiences and thus become a factor, which influences group decision and negotiation. This was the topic of Springer's collected volume on Emotion in Group Decision and Negotiation, 2015, edited by Bilyana Martinovski. We therefore invited research, which explore further the following questions: How do emotions become inter-subjective? How do emotions influence the negotiation process/outcome, and vice versa? What are the neural conditions and functions of emotions in group decision and negotiation? What are the communicative means for the realization of emotions? How are emotions related to ideology and identity? How do different types of media affect emotions in group decision and negotiation contexts? Are there different cultural aspects of emotions? What methods could be used for the study of emotions in group decision and negotiation?

One peer-reviewed full paper entitled 'Effects of Pre-Negotiation Behavior on the Subsequent Episode' was accepted for this volume. It is authored by scholars from University of Potsdam, Marie-Christin Weber and Uta Herbst, in collaboration with scholars from the University of Hohenheim, Marc Schmidt and Markus Voeth. The paper explores how buyer strategies applied to online pre-negotiations affect the attitude of the supplier party and the outcome of negotiation. It defines two strategies, 'power' verses 'visionary'. The paper associates power strategy with negative emotions expressed by speech acts, such as threats and bluffs. It is described as 'imposing strict requirements to which suppliers have to respond in order to become selected for the main negotiation'. Visionary strategy is associated with positive emotions and applied when 'buyers try to enthuse the counterpart in such a way that the supplier adjusts his goals in order to become part of the project'. The authors expect that buyer's use of visionary strategy, expressed by speech acts, such as compliments, in the pre-negotiation stage would result in supplier concessions during the negotiation. The simulated experiment data are gathered through a chat-based online negotiation tool and consist of 100 professionals matched with 10 trained student partners. Their task is to identify a suitable IT-system supplier for the creation of a self-driving car. Suppliers are interviewed on their aspiration for the upcoming main negotiation. Buyers, i.e. the students, rated their counterpart's behavior on items such as 'fact-based' or 'relationship-oriented' behavior. The study concludes that the authors' expectations are moderately validated: a visionary strategy applied to online chat-based pre-negotiations correlates with a positive impact on the main negotiation. The visionary strategy reaches a higher number of agreements than the power strategy. The group exposed to the visionary strategy reveals a higher willingness to concede than the power-related group.

Further research on emotions in pre-negotiations may explore relation to cheap talk and validity of findings in face-to-face settings, authentic settings and different types of negotiations. Positive speech acts such as compliments may very well be part of a diplomatic power play. Power behavior might be more effective in international interstate diplomacy negotiations than visionary behavior. If diplomats would reveal vision on the endgame the whole process may stop right away as different member states might agree on short-term decisions but not at all have the same long-term vision.

In sum, GDN2017's stream on emotions in group decision and negotiation develops novel models on how communication influences emotions and how emotions affect the process and outcome of negotiations. It explores different types of negotiation and negotiation stages and applies variety of methods for analysis while it focuses on perception of emotions in negotiations. Future research on emotions in group decision and negotiation would gain from inter-disciplinary collaboration, including neuro-science, economics, linguistics, computer science, and anthropology.

Effects of Pre-negotiation Behavior on the Subsequent Episode

Marie-Christin Weber[1]([⊠]), Marc Schmidt[2], Uta Herbst[1], and Markus Voeth[2]

[1] Department of Marketing II, University of Potsdam, August-Bebel-Straße 89,
14482 Potsdam, Germany
marie-christin.weber@uni-potsdam.de

[2] Department of Marketing and Business Development, University of Hohenheim,
Fruwirthstraße 32, 70599 Stuttgart, Germany
schmidt.marc@uni-hohenheim.de

Abstract. Research has dealt many times with behavior as success factor for negotiation outcome. So far, these factors are limited to a specific negotiation situation while no insights into a negotiator's most recommendable behavior for successive episodes exist. Accordingly, in this work, the effect of pre-negotiation behavior on the main negotiation episode is examined through a buyer-seller negotiation experiment. Records of online chats were investigated by means of content analysis and combined with questionnaire results in order to reveal behavioral patterns in the pre-negotiation and their consequences onto the main negotiation. Two main behavioral streams have been identified: visionary behavior that emphasizes future collaboration in a positive setting and power-related behavior that rather discloses dominating and imposing elements. This study found that visionary behavior in the pre-negotiation episode leads to more success than a power-related behavior in terms of the general agreement on a negotiation's conduction and the opponent's concession attitude.

Keywords: Pre-negotiation · Buyer-supplier negotiation · Power · Vision · Communication

1 Introduction

Negotiations oftentimes do not only consist of one single interaction where parties meet and try to come to an agreement, but are rather often divided into several negotiation episodes that are built on one another [1, 2]. Consequently, many negotiation parties meet several times (e.g. pre- and main negotiation) until they reach a final agreement. In the case of buyer-seller negotiations this is due to the fact that buyers and suppliers mostly do not only have to agree on the price issue but also on several complex as well as highly technical specifications [3, 4]. Whereas the price agreement usually is settled well at the end of the negotiation process (main negotiation episode), pre-negotiation episodes are usually used from the buyer's point of view to find out whether the supplier brings along sufficient competencies to meet the technical specifications and whether he is capable of keeping up with time and financial requirements given by the buyers.

© Springer International Publishing AG 2017
M. Schoop and D.M. Kilgour (Eds.): GDN 2017, LNBIP 293, pp. 91–101, 2017.
DOI: 10.1007/978-3-319-63546-0_7

Thus, pre-negotiations oftentimes resemble pre-selection meetings on the basis of which one may find out whether a future cooperation is conceivable [5].

However, in addition to a simple clarification of whether or not to further negotiate with a certain supplier, buyers could also use the pre-negotiation episode strategically in order to better achieve their interests in the main negotiation. In this context, both a powerful as well as a visionary strategy are feasible [6, 7]. The former shapes the setting in a negative way by imposing strict requirements to which suppliers have to respond in order to become selected for the main negotiation. In contrast, in the case of the visionary strategy buyers try to enthuse the counterpart in such a way that the supplier – rather voluntarily, because intrinsically motivated – adjusts his goals in order to become part of the project. Organizational studies have shown that very high intrinsic incentives can compensate lower extrinsic aspects such as payment [8, 9]. In the context of buyer-seller negotiations, it can be assumed that if a selling firm develops high levels of enthusiasm for a project, it will reduce its extrinsic reward expectation. More concretely, it can be suspected that an intrinsically motivated supplier will make higher concessions in the main negotiation. These assumptions are further supported by the theory of negotiation framing. Based on prospect theory, framing theory states a better negotiation outcome when the negotiation setting is introduced in a positive way [10]. In contrast, if entering a negotiation with a negative frame, outcomes will decrease.

Much research exists dealing with behavior within the main negotiation and its effect on the outcome. But, to the best of our knowledge, research did not yet empirically investigate the effect of behavior from one episode on behavior in a subsequent one (e.g. from pre-negotiation episode on the main negotiation episode). Against this background, the goal of this study is to examine if and how behavior in the pre-negotiation episode can influence a situation in such a way that it shapes the opponent's attitude towards a positive or negative frame in the main negotiation episode (power vs. visionary strategy). Therefore, we start with a short break down of existing insights into prospect theory and negotiation framing and thereby differentiate between a visionary and a power-related behavior. In combination with management theory we apply the two behavioral patterns on the pre-negotiation episode and therewith establish our hypotheses. We then present our experiment and the method of content analysis. Afterwards, we report the results and conclude with new insights for both academia and practice.

2   Theoretical Background

Interaction between parties in buyer-seller negotiations is not limited to the main nego-tiation, but parties rather communicate at least once in the pre-negotiation episode or sometimes even on a regular basis. In general, a negotiation can be understood as a multi-episode process that begins with the pre-negotiation episode [11]. This episode is defined as a process of communication due to some kind of conflict ending with the agreement upon the continuation of communication in form of a negotiation or with the abortion of the parties' interaction [12]. Hence, communication in the pre-negotiation episode represents a main trigger for a main negotiation's realization. In other words, the pre-negotiation episode describes every interaction between the parties related to and taking place before the main negotiation. Parties might exchange information and attitudes that concern the negotiation during this episode, but are not trying to reach an agreement yet.

Several insights are gained considering a self-imposed negotiation frame in the pre-negotiation episode and its influence on the successive course of action [13]. Prospect theory, as basis for framing theory, illustrates that people evaluate a situation as positive or negative based on the designation of either losses or profits they are facing [14, 15]. In the context of business negotiations, gains and damages are oftentimes expressed in terms of financial standings or opportunities of growth and knowledge access. Depending on if a party evaluates possible negotiation outcomes as positive or negative (and thereby frames the whole setting in the respective direction) the results differ: while emphasizing that any concession will generate financial losses for the company leads to a less risky behavior and results in less concessions and less settlements, highlighting the gains of a successful negotiation conducts the opposite effect [10]. We already know that a frame affects behavior [16] and more concretely that a positive frame demonstrably leads to more settlements than a negative frame [17]. Neale and Bazerman [10] further found that a positive frame results in better outcomes than a negative one due to the counterpart's higher willingness to concede in positive settings. Consequently, a person's attitude when entering a negotiation is supposed to have a great effect on the negotiation. In this context, one could argue that a positive attitude favors the individual outcome. However, what we do not know is how our behavior can influence a situation in such a way that we shape the opponent's attitude towards a positive or negative frame.

Within a negotiation, behavior finds its expression pre-dominantly in communication which in turn discloses diverse emotions. Emotions have the power to affect a negotiation's process and thereby its outcome and success [e.g. 18, 19]. They are revealed through positive or negative communicational elements such as threats, compliments, or rejections. As prospect theory explains, the words used have the power to influence a situation's atmosphere. When classifying behavioral characteristics into a positive or rather negative frame, a positive expression can be equalized with the visionary manifestation of future collaboration and a comfortable relationship between the parties whereas negative elements (e.g. threats and bluffs) resemble expressions of a powerful behavior [20]. Thus, in the visionary condition the party emphasizes the gain of future collaboration, while in the power condition losses are imposed by strict requirements.

In contrast to the beneficial effect of a positive frame, scholars found that within the main negotiation powerful negotiators reach better outcomes and concede less while their counterpart is even more willing to make concessions [e.g. 21–23]. This is due to the intimidating effect of powerful behavior: as the counterpart perceives this behavior as an expression of high limits, he rather makes higher concessions than losing the whole argument [23]. Although the demonstration of power might be successful in the main negotiation, this might not necessarily apply for the pre-negotiation episode as well. In the pre-negotiation episode parties are not supposed to reach final agreements on the negotiation issues. Rather, this episode is used to pave the way for future collaboration and the following main negotiation. Hence, the primary aim is the agreement on a negotiation's conduction. As negotiation framing states that a positive frame favors settlements in the main negotiation, we can assume that a positive frame could also benefit an agreement on the transition from the pre- to the main negotiation episode. However, the intimidating effect of power-related behavior could cause an abortion of further communication instead of its continuation in the shape of a negotiation. Then, visionary behavior would represent a more successful strategy than the demonstration of power.

H1:   Visionary behavior in the pre-negotiation episode leads to more agreements on entering a negotiation than the demonstration of power.

Negotiations are often not only driven by economic but rather by intrinsic means [13]. Indeed, insights of organizational and psychological behavioral approaches in terms of motivation underline the importance of intrinsic incentives [e.g. 24–26]. While people mostly get compensated and motivated for work by tangible rewards such as money, one can also be motivated by intrinsic factors like joy or satisfaction [27]. This construct does not only apply for job-related situations, but can rather be found in very different contexts [28]. Gagné and Deci [28] introduced a form of intrinsic motivation called integrated regulation. It is based on autonomous motivation stemming from consistency with one's own values and goals. Furthermore, Gagné and Deci [28] state that motivation has greater impact on satisfaction and performance when it is not imposed by external factors but rather stems from a person's convections. Hence, intrinsic motivation can be more important than extrinsic incentives so that very high intrinsic stimuli might even compensate lower extrinsic rewards.

Usually a supplier receives a compensation for the work and goods he delivers in form of monetary means. The height of this sum represents a main issue of the negotiation process and is a decisive decision criterion for or against a collaboration. Oftentimes other aspects of compensation for the supplier's motivation are not considered. Transferring the insights of motivational theory to the supplier selection process, there are different ways to satisfy a supplier. Thus, a supplier becomes not only motivated by the price he receives in exchange for his work, but also in an intrinsic way, when a project itself enthuses him in such a way due to personal values or future opportunities that it generates more satisfaction and joy than money would do. As intrinsic motivation has the potential of having a stronger impact than extrinsic incentives, one can assume that, if a supplier wants to become part of a project for intrinsic reasons, he might even lower his external reward expectation. The intrinsic rewards would then cover the lack in extrinsic outcomes. A buyer can call a supplier's attention for intrinsic rewards by emphasizing positive aspects of a future collaboration. That can be achieved by pointing out benefits of the project and by giving the impression of a comfortable atmosphere through positive emotions or positive relation messages. Hence, a buyer generates a positive frame for the upcoming negotiation by developing a vision. Consequently, the supplier might be enthused for an upcoming project in such a way that he will rather make higher concessions in the negotiation than losing the deal. So, if he has been set in a positive frame that emphasizes a positive relation and benefits of future cooperation, he might enter the main negotiation with a higher willingness to concede.

H2:   Counterparts that are triggered by visionary behavior in the pre-negotiation episode are more likely to make concessions than those that are confronted with power.

3   Method

Participants. A total of 100 professional buyers from an automotive supplier took part in a company-wide negotiation competition. Each of them was randomly assigned to one of 10 trained research assistants, creating a total of 100 dyads. Research assistants were trained in negotiation for about a year and were advised to react in the same way by using a standardized statement data base. By doing so, we standardized the negotiation counterpart as far as possible in order to make the behavior of the participants comparable. Herbst and Schwartz [29] found that students can be trained regarding their negotiation behavior and then perform in the same way as professional negotiators. Participants' motivation for a successful attendance and realistic behavior was created through a competition in which the best three negotiators as well as one randomly assigned participant won prizes. Participation in the competition was voluntarily. Individual performance was confidential and it was assured that personal results would not be passed on to the company.

Procedure and task. In order to examine our hypotheses, we applied a multi-method approach. It combined an experiment simulating the pre-negotiation episode with a questionnaire that was used to assess expected behavior and results of the main negotiation episode. In the experiment, participants had to conduct a pre-negotiation buyer-supplier case regarding an automotive component system via an online chat tool. The case was designed especially for this experiment based on six extensive expert interviews about the comprehensive supplier selection process. Thereby, we guaranteed a case that simulates the pre-negotiation episode in a realistic way. One week prior to the negotiation all participants obtained extensive role descriptions and background information giving them enough time to familiarize with the issues and prepare the negotiation in order to create a situation close to reality. While the role of the supplier was executed by trained students, all participants represented buyers of an automotive company responsible for the supplier selection process. Their task was to identify a suitable IT-system supplier for the creation of a self-driving car. During an online chat, participants had to determine if the supplier is able to implement technical requirements and had to discuss four items with integrative potential in terms of the cooperation's realization. In order to encourage participants to negotiate as they would do in a real life business context, they were told that their counterpart was embodied by a professional negotiator. If the pre-negotiation talk was successful and the parties wanted to enter the negotiation, they had to sign a sheet of cooperation and determine framework conditions for each issue. The goal of each supplier was to already agree on favorable conditions, meaning conditions that are at low cost for the buyer. Further, they were told that the final negotiation would take place the next day to make sure that they put themselves in the situation of the pre-negotiation episode and do not interchange it with the main negotiation. Each dyad had a time slot of 45 min which automatically closed five minutes after the time has elapsed. Subsequently to the negotiation task, we interviewed suppliers on their aspiration for the upcoming main negotiation in regard to each negotiation object. Moreover, the students had to rate their counterpart's behavior on a 3-point scale (agree, not sure, disagree) regarding items such as "fact-based" or "relationship-oriented". Furthermore, a 5-point Likert scale was used to measure the supplier's willingness to concede regarding each single item before and after the experiment anchored by 1 (strongly disagree) and 5 (strongly agree).

Analysis. In order to measure the impact of behavior in the pre-negotiation episode on the main negotiation's outcomes, we first coded all chat records of the conducted experiment and analyzed them by means of content analysis. Based on recent studies on emotions in negotiation [30, 31], we used fourteen items and counted statements that included them. As we wanted to determine the effect of power and visionary factors, the items have been assigned to either one of the categories. Power-related elements represent those communication items that use any kind of pressure on the counterpart to fulfil specific requirements for being considered as possible cooperation partner [20]. Those elements are threats, rejections, negative reactions, promises ("if you..., then..."), reference to BATNA or reservation price, and negative emotions. Meanwhile, visionary arguments were coded by the items positive relation message, compliments, small talk, emoticons, enthusiasm for the topic, enthusiasm for the project, and positive emotions. All these communication items are used to frame the negotiation in a positive way in order to make the counterpart wanting to become part of the project. The according questionnaires were analyzed by means over the categories.

Reliability assessment. Every chat was content analyzed by two researchers and randomly counterchecked by a third coder. Inter-coder reliability was measured with Guetzkow's U = .0071 which represents a good to very good fit [33]. Moreover, the index of coterminability also reached a good response rate with 73% [34]. Reliability over all categories was measured with .767, also representing a satisfactory result [35]. As only marginal variance has been detected, we can assume inter-rater reliability.

Validity assessment. For the items in our coding scheme we relied on existing schemes already used in negotiation research that measure communicational elements [30–32]. As the items' validity has been proven in the past, we can assume their validity. Moreover, we pre-tested all items and thereby could guarantee face validity. We added the items "enthusiasm for topic" and "enthusiasm for project", because they were of great interest to us. Again, we pre-tested those categories in advance in order to maintain validity under analysis extension.

4   Results

Of all participating buyers, a total of 98 negotiation dyads could be used for further analysis. Others needed to be excluded for reasons such as disregard of instructions or technical issues. We divided the sample into two groups: one that applied more power-related than visionary arguments $(N = 37)$ and another group of people that revealed more visionary than power-related communicational statements $(N = 56)$. The remaining five subjects did not report a significant difference of strategy usage and were therefore no longer investigated. We found that significantly more people that applied a visionary strategy reached an agreement *(n = 43 (77%), p < .01)* than people applying a power-related strategy *(n = 24 (65%))*. Thus, hypothesis 1 is supported (Table 1).

Table 1. Percentage of contracts reached regarding applied strategy

	Power	Visionary
Contract	64.86%	76.79%
No contract	35.14%	23.21%
N	37	56

Next, we examined the effect that the application of visionary or power-related arguments has on the counterpart's willingness to make concessions. Correlation between single items as well as the categories power and vision with the counterpart's willingness to make concessions have been tested with Spearman's correlation coefficient. While no correlation between the use of powerful arguments and the counterpart's willingness to make concessions was shown *(r = .055, p > .1)*, we found a slight, but significant correlation between the number of visionary arguments used and the counterpart's willingness to make concessions *(r = .151, p < .07)*. Hence, we can notice a coherence between the number of positive motivational incentives used and the opponent's concession attitude (Table 2).

Table 2. Correlation between strategy and counterpart's willingness to concede

	Power	Visionary
Correlation coefficient (Spearman)	.055	.151
p (one-sided)	.296	.069*
N	37	56

* statistically significant (p < .1)

A closer look at the single items of each strategy reveals that the negative correlation with the counterpart's willingness to concede is significantly guided by the items rejections (r = −.166, p < .09) and negative reactions (r = −.255, p < .04). However, lies and references to BATNA or reservation price show a positive correlation, but at no significant level (Table 3). On the side of the visionary statements the items emoticons (r = .136, p < .09) and positive emotions (r = .312, p < .01) show the greatest effect on the counterpart's willingness to make concession. The items enthusiasm for project and positive relation message also reveal a positive correlation (r = .108, r = .149), but at no significant level (Table 4).

Hence, we found that on the one hand the group of people applying a visionary strategy reached a higher number of agreements with their counterpart than the group of people using mainly power-related arguments. Moreover, opponents of people that can be assigned to the visionary strategy revealed a higher willingness to concede than the power-related group. Thus, the visionary strategy shows a moderately positive impact on the main negotiation episode. Hypothesis 2 is supported.

Table 3. Correlation coefficient of counterpart's willingness to concede and number of power-related arguments

Power		
Item	Corr. Coef. (score)	P-Value (one-sided)
Threats	−.045	.356
Rejection	−.166	.085*
Negative Reaction	−.225	.031**
Lies	.126	.150
Promises	.078	.259
Reference to BATNA or reservation price	.125	.151
Negative emotions	.044	.358

* statistically significant (p < .1)
** statistically significant (p < .05)

Table 4. Correlation coefficient of counterpart's willingness to concede and number of visionary arguments

Visionary		
Item	Corr. Coef. (score)	P-Value (one-sided)
Enthusiasm for Topic	−.062	.304
Enthusiasm for Project	.108	.187
Positive Relation Message	.149	.110
Compliments	−.110	.182
Small Talk	.059	.313
Emoticons	.136	.089*
Positive emotions	.312	.004**

* statistically significant (p < .1)
** statistically significant (p < .01)

5   Discussion

This study represents the first work dealing with behavioral effects of one episode on a subsequent negotiation episode. The goal was to examine if and how behavior can influence a situation in such a way that it shapes the opponent's attitude towards a positive or negative frame (power vs. visionary strategy). Therefore, we investigated the pre- and the main negotiation episode. We found that the pre-negotiation episode and especially the communication within this stage states an important part of the whole negotiation process. Framing theory has already proven that the specification of a party's self-imposed frame (positively or negatively) influences the outcome in the main negotiation. This study has now shown that one can also oppose a frame on a counterpart and thus, slightly influence his concession behavior. In contrast to the main negotiation, a successful outcome of the pre-negotiation episode cannot be achieved by powerful behavior. In this episode rather a positive and visionary communicational behavior represents a key factor of success: Negotiators that reveal a positive and vision-based behavior reach more often a settlement for upcoming negotiations and can also notice a higher willingness to concede of their opponent. In contrast, obvious power-based behavior traits such as threats, rejections, and negative reactions show a slightly negative correlation with the counterpart's concession attitude. On the other hand, power-related communicational behavior in terms of lies and references to BATNA or reservation price are not received, if noticed, in a negative way by the counterpart and neither show a negative correlation.

Implications. Results prove that previous episodes, here investigated by the pre-nego-tiation episode, have a significant impact on the subsequent negotiation process and can already influence the decision on whether to enter a negotiation. While a power-related attitude might frighten possible collaboration partners, one can enthuse their counterpart with a positive communication style. This behavior can pay off in the final price nego-tiation. Hence, the pre-negotiation episode should not only be used for preparation and planning of the main negotiation. In fact, any interaction in the pre-negotiation episode should also be carefully executed and strategically used. More concretely, negotiators should avoid using negative reactions and aggressive rejections in the pre-negotiation episode and should instead underline positive emotions. The aim should be to create a positive and visionary frame for the upcoming negotiation. Surely, many other aspects influence the negotiation process and should therefore be considered as well.

Limitations. Our study surely reveals some shortcomings. First, we only investigated the buyers' side. When considering successful behavior of the supplier in the pre-nego-tiation episode, results could be different. Second, the case was designed about a very specialized and innovative product that made up a great deal for the buyer's company. Hence, different results could be revealed when regarding a less specialized product that is of less importance to the company. Moreover, we exclusively simulated the pre-negotiation episode and checked the opponent's concession behavior with a question-naire. Hence, we do not have proof for the actual negotiation outcome and interdepend-ence between behavior in the pre- and main negotiation episode. Also, there might be more than one situation of interaction between the parties in the pre-negotiation episode that influences the upcoming process. Lastly, the experiment was conducted through a chat-based online negotiation tool. Especially when considering emotions the applica-tion of this tool has some shortcomings: we can only investigate verbal communication and even this is limited to subjective interpretation of the coders. An investigation of the same experiment but conducted face-to-face or through different media could result in different insights. Nevertheless, an analysis of communicational behavior will always be limited to a subjective evaluation.

Future Research. Although there always exists some form of interaction between nego-tiation parties in the pre-negotiation episode, still very little is known about this episode's effect on the main negotiation. This study represents one of the first investigations to take a closer look at this topic. It can be expected that there will emerge many more studies regarding the above discussed topic in the future. We investigated the correlation between the way of communication and the resulting acceptance of concession making. As we only focused on buyer's behavior, it would be interesting to investigate the most favorable behavior for suppliers in contrast. Moreover, one could replicate our study in a political or diplomatic setting. While the visionary behavior might be successful in a business context, this might not necessarily apply for those areas as well. In fact, there, the presentation of power could be fundamental while the presentation of a long-term vision could even weaker one's position as politics in general are characterized by power games. Other fields of interest are the effects of communication in the pre-negotiation episode regarding different media used or the number of contact points in advance to the negotiation process. Also, research should investigate the interaction of behavior in the pre- and main negotiation episode. So far, we assume, based on our findings and theory, that a change from visionary behavior in the pre-negotiation episode to a more powerful style in the main negotiation represents the most successful strategy, but we do not have empirical support for this assumption. Moreover, as our case study was based on a specialized product, it would be very interesting to investigate the effect of the pre-negotiation episode when regarding a commodity product in order to see if in this case the pre-negotiation episode shows any effect as well. Finally, our experiment was designed in an observatory way. We did not impose any manipulations, because we first wanted to examine if a behavioral classification of the two groups (visionary and power-related behavior) can be noticed. As this indication was proven, the study should be replicated with two manipulations: one group that is told to use arguments that demonstrate power and another group that tries to influence their counterpart by motivational incentives. Such a study design should lead to results that are even more significant.

References

1. Harinck, F., de Dreu, C.K.W.: Take a Break! Or not? The impact of mindsets during breaks on negotiation processes and outcomes. J. Exp. Soc. Psychol. **44**, 397–404 (2008)
2. Harinck, F., de Dreu, C.K.W.: When does taking a break help in negotiations? The influence of breaks and social motivation on negotiation processes and outcomes. Negot. Confl. Manage. Res. **4**(1), 33–46 (2011)
3. Van der Valk, W., Wynstra, F.: Buyer-supplier interaction in business-to-business services: a typology test using case research. J. Purchasing Supply Manage. **18**(3), 137–147 (2012)
4. Thomas, S.P., Thomas, R.W., Manrodt, K.B., Rutner, S.M.: An experimental test of negotiation strategy effects on knowledge sharing intentions in buyer-supplier relationships. J. Supply Chain Manage. **49**(2), 96–113 (2013)
5. Kim, M., Boo, S.: Understanding supplier selection criteria: meeting planners' approaches to selecting and maintaining suppliers. J. Travel Tourism Mark. **27**, 507–518 (2010)
6. Varadarajan, R.: Strategic marketing and marketing strategy: domain, definition, fundamental issues and foundational premises. J. Acad. Mark. Sci. **38**, 119–140 (2010)
7. Steensma, H., van Milligan, F.: Bases of power, procedural justice and outcomes of mergers: the power and visionary factors of influence tactics. J. Collective Negot. **30**(2), 113–134 (2003)
8. Herzberg, F., Mausner, B., Snyderman, B.: The Motivation to Work. Wiley, New York (1959)
9. Petrescu, A.L., Simmons, R.: Human resource management practices and worker's job satisfaction. Int. J. Manpower **29**(7), 651–667 (2008)

10. Neale, M.A., Bazerman, M.H.: The effects of framing and negotiator overconfidence on bargaining behaviors and outcomes. Acad. Manag. J. **28**(1), 34–49 (1985)
11. Crump, L.: Analyzing complex negotiations. Negot. J. **31**(2), 131–153 (2015)
12. Schiff, A.: Pre-negotiation and its limits in ethno-national conflicts: a systematic analysis of process and outcomes in the cyprus negotiations. Int. Negot. **13**, 387–412 (2008)
13. Hunt, C.S., Kernan, M.C.: Framing negotiations in effective terms: methodological and preliminary theoretical findings. Int. J. Confl. Manage. **16**(2), 128–156 (2005)
14. Kahneman, D., Tversky, A.: Prospect theory: an analysis of decisions under risk. Econometrica **47**, 263–291 (1979)
15. Tversky, A., Kahneman, D.: The framing of decisions and the psychology of choice. Science **211**, 453–463 (1981)
16. Pruitt, D.G., Carnevale, P.J.: Negotiation in Social Conflict. Open University Press, Buckingham (1993)
17. Crawford, V.P.: On compulsory arbitration schemes. J. Polit. Econ. **87**(1), 131–159 (1979)
18. Olekalns, M., Druckman, D.: With feeling: how emotions shape negotiations. Negot. J. **30**(4), 455–478 (2014)
19. Kopelman, S., Rosette, A.S., Thompson, L.: The three faces of eve: strategic displays of positive, negative, and neutral emotions in negotiations. Organ. Behav. Hum. Decis. Process. **99**, 81–101 (2006)
20. Lawler, E.J.: Power processes in bargaining. Sociol. Q. **33**, 17–34 (1992)
21. De Dreu, C.K.W.: Coercive power and concession making in bilateral negotiation. J. Conflict Resolut. **39**, 646–670 (1995)
22. Giebels, E., De Dreu, C.K.W., Van de Vliert, E.: Interdependence in negotiation: effects of exit options and social motive on distributive and integrative negotiation. Eur. J. Soc. Psychol. **30**, 255–272 (2000)
23. Van Kleef, G., De Dreu, C.K.W., Pietroni, D., Manstead, A.S.R.: Power and emotion in negotiation: power moderates the interpersonal effects of anger and happiness on concession making. J. Soc. Psychol. **36**, 557–581 (2006)
24. Maslow, A.H.: Motivation and Personality. Harper & Row, New York (1954)
25. Herzberg, F.: Work and the Nature of Men. World Publishing, Cleveland (1966)
26. Alderfer, C.P.: Existence, Relatedness, and Growth. Free Press, New York (1972)
27. Deci, E.L.: Intrinsic motivation, extrinsic reinforcement, and inequity. J. Pers. Soc. Psychol. **22**, 113–120 (1972)
28. Gagné, M., Deci, E.L.: Self-determination theory and work motivation. J. Organ. Behav. **26**, 331–362 (2005)
29. Herbst, U., Schwarz, S.: How valid is negotiation research based on student sample groups? new insights into a long-standing controversy. Negotiation Journal **27**(2), 147–170 (2011)
30. O'Connor, K., Adams, A.A.: What novices think about negotiation: a content analysis of scripts. Negot. J. **15**, 135–147 (1999)
31. Adair, W.L., Loewenstein, J.: Talking it through: communication sequences in negotiation. In: Olekalns, M., Adair, W.L. (eds.) Handbook of Research in Negotiation, pp. 3–23. Edward Elgar, New York (2013)
32. Pesic, M.: Emotionen in Verhandlungen. Hamburg, Dr. Kovač (2009)
33. Guetzkow, H.: Unitizing and categorizing problems in coding qualitative data. J. Clin. Psychol. **6**(1), 47–58 (1950)
34. Angelmar, R., Stern, L.W.: Development of a content analytic system for analysis of bargaining communication in marketing. J. Market Res. **15**(1), 93–102 (1978)
35. Holsti, O.R.: Content Analysis for the Social Sciences and Humanities. Addison-Wesley, New York (1969)

Negotiation Support Systems and Studies

University of Hohenheim

Stream Introduction: Negotiation Support Systems and Studies

Sabine T. Koeszegi[1] and Gregory E. Kersten[2]

[1] Institute of Management Science, Vienna University of Technology (TU
WIEN), Theresianumgasse 27, 1040 Vienna, Austria
Sabine.Koeszegi@tuwien.ac.at
[2] J. Molson School of Business, Concordia University, 1455 De Maisonneuve
Blvd. W., Montreal PQ, Canada
gregory.kersten@concordia.ca

1 Introduction

Over the last years, electronic negotiations have become an accepted form of conducting business transactions. They complement and in many situations, replace the traditional face-to-face negotiations. While negotiators in organizational negotiation processes might use general communication systems such as email or Skype, there are also systems that are more specifically targeted at e-negotiations. Because business negotiations are a component of business exchanges, many of these systems have been embedded in the supply chain management software platforms offered by IBM, Oracle, SAP and other firms. There are also stand-alone negotiation support systems (NSSs), developed to support communication, decision making, document management, and/or conflict resolution in social and economic contexts.

Over the past decades, we have seen sophisticated NSSs that provide both holistic and/or analytic support for all negotiation activities. In addition, negotiation software agents have been designed either to engage in automated negotiations or to aid human negotiators in their activities. Both the systems and the agents have been tested in various experiments and have been shown to improve the process and the outcome. Despite these achievements, scholars continuously point to issues that need to get more attention in order to improve effectiveness of negotiation support. The papers in this section provide examples of how to improve next generation negotiation support. The researchers, developers, and practitioners who design and develop NSSs, study their use in the laboratories and in the field and provide valuable insights into complex processes on various levels. While the first two papers address the design of systems to facilitate the modeling of the negotiation process and the negotiation problem, the following two papers deal with the facilitation of communication, interaction, and convergence towards agreements in NSSs-facilitated negotiations.

2 Overview

The first paper by William W. Baber introduces a bird's eye level phase model of negotiation processes based on a comprehensive and careful review of phase models suggested in research and practice. Although the suggested guideline for the sequence of macro phases in negotiations is not targeting negotiation support, it will undoubtedly facilitate the modeling of processes in electronic negotiation support and thus constitutes a very welcome and valuable contribution to the field.

In the second paper, Annika Lenz and Mareike Schoop develop a valuable framework for the analysis of the structure of decision problems in requirements negotiations. In order to realize the full potential of negotiation support systems in these types of negotiations, the support system has to be adapted to the structural specificities of the decision problem. Their qualitative-descriptive analysis approach results in the provision of matrix of possible scenarios of decision problems that can be utilized to leverage the effectiveness of negotiation support in requirement negotiation settings.

Previous literature has shown that the effect of negotiation support systems on processes and outcomes is moderated by various variables, amongst others by the cultural background of their users. The paper by Nil-Jana Akpinar, Simon Alfano, Gregory Kersten and Bo Yu focusses on the role of culture as a moderator in computer-mediated communication processes. In their analysis of empirical data gathered from online negotiations, they show that emotional contagion, measured by the reciprocation of the language sentiment in written text messages, is less pronounced in intercultural negotiations as compared to intracultural negotiations. They also observe that negotiators who initiated conversation, which need not include a formal offer, achieved greater payoff than the responding negotiators.

Finally, Gabriel Guckenbiehl and Tobias Buer develop and test a mechanism to overcome deadlocks and to facilitate convergence towards agreements among automated agents in complex but standardized contract negotiations. They show in computation experiments that compensation payments that are based on the Nucleolus method from cooperative game theory and suggested by a mediator improve negotiation results, both in terms of quality and time compared to automated negotiation mechanisms without compensation payments.

A Lifecycle Macro Phase Model for Negotiation

William W. Baber[(✉)] [iD]

Kyoto University, Kyoto, Japan
baber@gsm.kyoto-u.ac.jp

Abstract. Existing models of negotiation as a process are incomplete and do not show an overall, end to end process. The phases of existing models have not been clearly defined by identifying their boundaries. After reviewing contributions of existing models, the paper identifies phases, clarifies their boundaries, and proposes a bird's eye level model supported by examples in academic literature and public sources. Although an ideal model, it is a guideline and not a strict prescription for success. The proposed model contributes to theory around negotiation by providing a clarifying look at the overall sequence of macro phases in negotiation. With the model, academics and practitioners have a unified starting point for monitoring, communicating, and further developing negotiation models.

Keywords: Negotiation · Phase model · Conflict resolution · Process model

1 Introduction

Negotiation is not only a vital business interaction, it has become a field of academic study crossing studies such as Management [1–3], Psychology [4, 5], International Business [6], Law [7], and International Relations [8], among others. Negotiation has been theorized variously, as dimensions [9], DNA [10], teams [11], and jazz [12]. Process is also a way to view negotiation. The importance of process to negotiation has been identified by academia [13–17]. Before mapping negotiation processes at detailed, disaggregated levels, however, the total lifecycle of negotiation must be considered and modeled in terms of its macro phases. Various process models are reviewed in this paper to ask (1) whether evidence for macro phases from inception to completion of a negotiation can be found, and (2) whether the boundaries of those phases can be determined.

Current negotiation models lack completeness as they may exclude activities before or after the main negotiation interactions. Further, they may lack features such as feedback loops which return negotiators to previous phases with new information and decision gates to quit or continue. Such features would make models more accurate and usable to theoreticians, educators, and practitioners of negotiation who will benefit by gaining new theory building tools, teaching insights, and best practices. This article draws on documented negotiations to contribute to the conversation about negotiation phases the following: evidence of phases, their characteristics, boundaries of phases, transition across the model including feedback loops, and a full macro phase process model of the negotiation lifecycle.

© Springer International Publishing AG 2017
M. Schoop and D.M. Kilgour (Eds.): GDN 2017, LNBIP 293, pp. 107–119, 2017.
DOI: 10.1007/978-3-319-63546-0_8

2   Review of Existing Phasic Negotiation Models

The literature about negotiation has considered the process of negotiation, offering a variety of phase based models explicated in graphic or written form since the 1960s. Phases, also referred to as stages, are an appropriate approach for understanding negotiation because negotiation is a sequence of activities that progresses over time with differentiation among major activities that segment the end to end negotiation [15]. Additionally, Holmes [15] refers to Abbott [18] in pointing out that a phase model, if accurate, allows detection of a current phase and prediction of coming actions. Such phases represent large scale structures of the overall negotiation and are termed macro phases in this article in order to distinguish them from meso and micro level phases; smaller episodic [15] or sequence based [19] structures. Identifying phases at any level requires criteria. Thus efforts have been made to identify meso phase structures by sequence of activity [20], and micro phase structures through punctuation of sequences of interactions, for example, by breakpoints [19, 21] or turning points [22]. These however do not always indicate change to a new macro phase. Previous work has used text analysis to describe transitions among topics and strategies [19, 23]. Analysis of interactions has been used to separate meso and micro phases based on structural dissimilarities in communication acts identifying phases that recur in various kinds of face to face negotiations [16, 24]; these smaller phases however reside within an aggregated, macro phase which arches over the interaction of the parties. While the analysis of meso and micro phases sheds light on the workings of the macro phase where parties interact, it cannot reveal the nature of yet other macro phases in the end to end negotiation that extend from early considerations of the environment in which the deal and parties exist all the way through final phases when agreements are implemented and parties take stock of their performance. Rather than identifying the boundaries of the macro phase through statistical or text analyses, this study defines boundaries by outputs such as artifacts and documents [25, 26], major shifts in focus [27], the content of communications [28].

High level phasic segmentation of negotiations may not reflect real life because it is messier than linear models where there is no communication or reverse movement among phases [19]. Although these authors criticized phase models as simplistically progressive, they nonetheless use terms like "forward progress" [19] revealing at least some agreement with Holmes [15] that progress is inherent in a negotiation.

Table 1 below provides an overview of macro phasic models of negotiation including recent and older models that are still relevant in academic writing. Table 1 excludes models that handle only meso or micro phases.

Summary of Existing Models
The models compared in Table 1 are widely disparate, nonetheless all explicitly depict negotiation as a series of steps processing to a conclusion. None of the phasic models reviewed combine all the necessary features and scope of an overall process model for negotiation. Some lack phases at the beginning or end, some focus on phases below the macro phase level, some are strictly linear in sequence and lack feedback loops, others lack decision points to proceed, return or stop. These omissions may stem from authors

creating descriptive models common to their industry or activity such as value creation late in the sequence [7] or no preparation phase in hostage negotiations [43]. The proposed model has all of these features and intends to provide a symmetrically prescriptive [44] model applicable to many kinds of negotiation as a guideline, not a straitjacket, because negotiations may be unique in content and context.

Table 1. Elements of current macro phase models

	Lifecycle	Feedback loops	Loop to start	Identifies actors	Decision gates
Douglas 3-Phase [29]	NA	NA	NA	IM	NA
Joint problem solving process [1]	NA	EX	EX	EX	NA
Morley Stephenson [30]	NA	NA	NA	IM	NA
Gulliver Processual [31]	PA	EX	NA	EX	NA
Zartman Berman [8]	PA	NA	NA	EX	NA
Brooks [32]	PA	EX	NA	NA	EX
Craver [7, 33]	PA	NA	NA	EX	NA
Heller et al. [34]	EX	PA	NA	NA	PA
Win-Win Spiral [35]	PA	IM	EX	NA	NA
Graphic Roadmap [36]	EX	NA	NA	EX	NA
Intentional Agent [37]	NA	EX	NA	NA	EX
MPARN [38]	PA	EX	NA	IM	NA
4-Phase Dance [39]	NA	NA	NA	NA	NA
Demirkan [40]	PA	EX	NA	IM	NA
CBI Mutual Gains [41]	PA	NA	NA	EX	NA
Five Stage [42]	PA	NA	IM	IM	NA
Fells et al. [20]	PA	NA	NA	NA	NA

Legend: EX = explicit; IM = implicit; NA = Not Appearing; PA = partial and explicit

3   Discussion and New Model

This paper proposes not only a pragmatic process model for understanding negotiation but also a rigorous one [45], that is, a model suitable for analysis because it can be reproduced, diagnosed, and improved based on the diagram. Processes may take input from other [46] organizations despite being set in action by only one organization [25], underlining that negotiation involves organizations as well as individual actors.

While taking on negotiation from the point of view process and sequence, this paper does not reje other approaches. Like other models, sequential phases are a cognitive attempt at sense making of human interactions. Dimensions have been proposed as a way to understand the changes in thinking and action of individuals as they maneuver towards their goal, operating now in one, now in another dimension though never divorced from any of them [9]. The advantage of the dimension viewpoint is that the actor can operate in any or all dimensions at the same time, avoiding the need to consider terms like forward, backward, progress, and movement. A model with parallel processes could be developed to emphasize that some processes may be active throughout a negotiation event. A process viewpoint nonetheless affords the freedom to move forward or back in the sequence through feedback loops which indicate that the step is to be partially or completely iterated with new information.

Variance models [15] intend to show cause and effect and may indirectly show sequence or process. Mechanisms such as moderating effects make this kind of analysis suitable for understanding decision making. However cause and effect are not always connected in a linear fashion and the related insights may or may not contradict or support understanding negotiation based on process. Facework seeks to explain negotiation choices and moves based on notions of managing respect and embarrassment [46]. Chinese negotiation has been described as having high impact from mechanisms around face, relationships, cognition, norms, and mores [47]. This kind of viewpoint provides a rich context for understanding negotiation, but may not generalize to non-Chinese cultures. Conversely, it remains to be seen if process models are germane to business practitioners working in a Chinese context.

The new model presented below does not disaggregate the macro phases at lower levels of process. The model presented here is an idealized, broadly prescriptive model developed from negotiation literature, observations of business practices, and experience. Diverse organizations will have specific needs and abilities or gaps in abilities, therefore this paper does not propose the model as suitable for all parties and situations. The simple model includes only sequential steps and not feedback loops, decision points, or actors. The rationale for the phases and their boundaries is explained below after further description of the overall model (Fig. 1).

Fig. 1. Simple negotiation model

A methodology to identify boundaries between phases is necessary. This article employs a combination of boundary defining evidence that is appropriate to the research questions [39]. Change of content has been used as an identifier of phases [39] where the intervals between phases are guided by theory and matched to empirical samples. In modelling of business processes, phase boundaries can be determined based on outputs such as documents and partially or fully completed products [25] or by change of interaction [27]. In projects, phase boundaries have been identified by artifacts such as agreements and signed plans or designs [26]. A project refers to an undertaking with a clear beginning and end with a unique outcome as a goal [48, 49]; a negotiation matches this definition as it is not a permanently on going operation and the intent is, for example, to come to agreements. In order to accomplish the identification of phase boundaries, negotiations published in academic sources and news media were reviewed for the presence of such boundary defining outputs and events. The Fig. 2 shows the macro phases and summarizes the phase boundary identifiers.

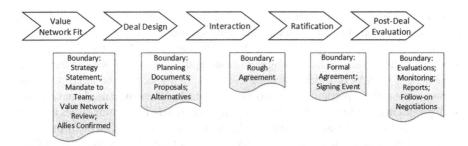

Fig. 2. Macro phase model with boundaries

Phase boundaries are a location for decision gates where a decision to continue, quit or return to an earlier phase, is made. Negotiations, despite fulfilling the definition of project, have not been modelled with the decision gates widely found in decision making processes in projectized organizations [50, 51]. The following Fig. 3 the landscape model with decision gates and feedback lines. The loop is closed with the last phase leading back to the start.

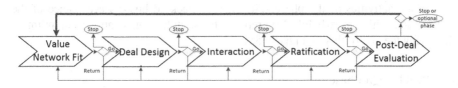

Fig. 3. Macro phase model with feedback loops

Based on the overview provided by the landscape model above, activities, aggregate activities, and outputs of each phase, are described below. Each macro phase of the landscape model is presented individually. The sub-activities of aggregate activities are not described or modeled; such work will have to wait for future research.

3.1 Value Network Fit Phase

The activities aggregated in this macro phase amount to a strategic review in which the organization's goals, allies, suppliers and competitors are considered in terms of the global value network and organizational strategy. The value network includes the partners, suppliers, competitors, and regulatory bodies that impact the company at a strategic level [52]. Not all deals are strategically important to the organization, therefore this phase may be omitted or shortened by the negotiators and their constituents. Transformative deals will necessarily require deeper review whereas mundane interactions will require little or no strategic review. This phase correlates roughly to the Search for Arena phase of Gulliver [31], the Diagnosis phase of Zartman and Berman [8], and the CBI [41] Prepare phase. In this model, however, the final decision makers and lead negotiator work with other strategic level staff to, for example, analyze stakeholders including allies and competitors, set broad goals and parameters, and assess the strategic fit of those goals up and down the value chain. Other actions aggregated here include identifying and contacting potential allies and counterparties, co-opting organizations, assessing likelihood of success, determining the Best Alternatives to Negotiated Agreement (BATNA) of each party, and initiating steps to strengthen one's BATNAs perhaps while weakening BATNAs of others. Doing these steps in advance of the interaction with counterparties allows participants to know or estimate the interwoven needs and interests of all parties in order to smooth problem solving and prepare for contingencies that may arise. The end of this phase is marked by the creation of overall goals, confirmation of allies, a written statement, or unwritten understanding of purpose. Ideally, there follows a considered decision to continue to the next phase, return to the Value Network Fit phase or abandon the project.

Examples: The 1997 negotiation involving UPS and a labor union highlights the Value Network Fit phase. Specifically, the union's methodical preparation gained the support of workers, politicians, and other stakeholders such as regular customers not normally part of a labor dispute [53, 54], exemplifying the strategic planning that characterizes the phase. Similarly, in the nine months before acquiring Autonomy in 2011, Hewlett-Packard sought a possible acquisition considering its allies and competitors carefully and conducted due diligence with the internal reports [55] that characterize the content and activities of the phase. The Hewlett-Packard board's acceptance of the potential acquisition in early July 2011 represents a strategic organizational commitment identifying the end of this first phase [55].

3.2 Deal Design Phase

This macro phase aggregates the activities of preparing offers and variations of packages to be offered as well as seeking new value creation opportunities. This phase corresponds partly to the dimension of the same name in the 3D negotiation model [9] but differs in that it includes team building like the Preliminary and Information phases [7], Planning phases and further information discovery [32]. The activity of setting negotiation goals found in this phase differs from the goal setting of the previous phrase the general goals are refined to be more specific. The main actors will be the lead negotiator, if the team is not self organized, the team intended to be in direct contact with the counterparties, and other team supports. At the same time, the lead negotiator is likely to liaise with the decision makers and strategic directors of the organization in order to synchronize goals and process. The end of this phase is defined by confirmation of those goals, revised information about BATNAs, allies, counterparties, and a set of potential offers and solutions [8]. The outputs of the phase may be in document form such as planning sheets, a project dashboard for the negotiation, statements about goals, and artifacts such as alternative plans or the briefs described by a diplomat looping through Deal Design and Interaction phases [12]. In the case of acquisition, due diligence documentation may be presented [56]. Finally, there is a decision gate to continue to the next phase, return to an earlier point in the Deal Design phase, return to the Value Network Fit phase, or abandon the project.

Examples: Labor negotiation transcriptions [29] show repackaging of offers by both sides when meeting separately from their counterparties. The parties cycle through joint interactions to private redesign efforts deal packages.

3.3 Interaction Phase

Despite the preparations in the first two phases, the Interaction macro phase, which aggregates the activities of the parties' communicating, means that new information, new solutions, new creativity, and new plans will inevitably arise. It is in the Interaction phase that the offers and ideas will be jointly developed until acceptable unless the negotiation is abandoned. The lead negotiator, if there is one, and the team members will be most active in this phase, however the leader and team may keep in contact with the final decision makers and call on the skills of other supporters as needed.

Discussion about the Interaction phase is well developed in the literature about negotiation, indeed some writers take negotiation to mean only the time during which parties are interacting [1, 38, 39]. However other writers see interactions as a phase, or phases, of a greater cycle, such as steps three through seven of MPARN [38], the Consensus Building Institute's Create and Distribute phases [41], and phases 3–5 of Gulliver's developmental model [31]. Gulliver's cyclical model [31] and Lopes' intelligent agents model [37] focus on the Interaction phase. MPARN, the cyclical model, and the interactive agents model show how iterative this phase is, as do the explanations around the CBI model [41]. Interactions, brief or long, can include proposals, and counter proposals [57, 58]. The end of the Interaction phase is marked by the cessation of these activities due to agreement or failure to agree. The agreement thus produced may be more or less formal, binding or not [20], and may not be in its final form [29].

Examples: An example of reaching the Interaction phase boundary is found in the transcripts of a labor-management negotiation. Just after general verbal agreement is reached the negotiation teams explicitly state that the agreement, despite enjoying their high confidence, is subject to ratification by the union membership [29]. The interaction phase in Hewlett-Packard's 2011 acquisition of Autonomy came when the two CEOs discussed financial packages in Deauville, France [55]. Finally there is a decision gate to continue to the next phase, return to an earlier point in the interaction phase or to an earlier phase, or cancel the project. In a sales negotiation between a UK equipment maker and a Finnish client [58], the decision gate was used once to return to the Deal Design phase from Interaction and then later to quit the negotiation.

3.4 Ratification Phase

This macro phase aggregates the activities of presenting outcomes to final decision makers such as superiors, the board, or peers, and attempting to gain their ratification as well as having legal experts finalize the language of an agreement. Such experts from all parties may work together in order to ensure smooth finalization or send the agreement back to the Interaction phase. While ratification is largely internal to any one party to the negotiation, it may mean a return to the Interaction phase with changes to be discussed and agreed with counterparties. The first action seen in the ratification phase is compiling the rough agreement, though this step may not be necessary for simpler documents in smaller organizations.

The end of this phase is marked by the output of a formalized agreement and perhaps ritual enactment, for example in a signing ceremony. Thereafter, a decision gate is entered to continue to the next phase, return to the Interaction phase or another previous phase, or abandon the project.

Example: The agreement jointly created in the Interaction phase must be confirmed by the final decision makers before and after legal write up, as in the case of the rough agreement reached by Hewlett-Packard and Autonomy in France that was later ratified by the board [55].

3.5 Evaluation and Monitoring Phase

This macro phase aggregates the following activities: implementation and execution of the agreement, monitoring of the execution, enforcement, strategic evaluation of the deal, evaluation of the team, formalization of learning points, renegotiation considerations, and relationship maintenance, among others. The strategic level evaluation of outcomes and satisfaction requires analysis by the strategic level managers. Evaluation of the team's performance may be conducted by the team, their observers, immediate leaders, and human resources staff. Monitoring of the execution of the agreement, as well as evaluation of the agreement, may include the counterparties or consultants and specialists in addition to the negotiation team. This phase includes the activities of Gulliver's [31] eighth phase, Execution of Outcome, as well as Brook's and Odiorne's eighth phase, Assessment and Performance Review [32] and is similar to the Follow Through phase of the CBI [41] model but does not include activities for developing enforcement mechanisms – those activities occur in the Interaction phase or at latest in the Ratification phase.

There are various potential formal or informal outputs to this phase: an improvement plan for the negotiation participants, intent to improve or break the relationship, evaluation of the counterparties' implementation of the agreement, on going monitoring, and commitments for follow up negotiations and renegotiations. Thereafter there is a decision gate to restart with a new deal with the partner(s) or exit the agreement. Thus the last phase is not the end of all activity; it may mean a return to the starting point of considering the strategic value fit with the partners or continuing to an optional phase to follow up with sub-agreements or renegotiation as the environment and project develop.

Examples: Post deal evaluation may lead to a dramatic exit from the agreement such as the firing of a top manager and legal challenges in the case of the Hewlett-Packard acquisition of Autonomy in 2011 [55]. A contract is itself a form of monitoring [59]. Self evaluation of negotiation teams is not seen as a standard practice, however some companies may self evaluate as part of knowledge management or good managerial and human resources practices.

3.6   Optional Follow-up Phase

This phase aggregates the activities of working out subordinate aspects of a deal and negotiating changes to a main agreement as the environment around the deal evolves. This phase is more likely to appear in the wake of agreements that are highly complex such as Service Level Agreements, multi decade agreements for resource development, and major infrastructure construction. Likewise, this phase is less likely to appear where deals are relatively simple or can be swiftly executed. Parties may enter this phase voluntarily in order to improve their outcomes in a post-settlement settlement [60]. The phase is therefore started, if at all, for reasons such as improving, completing, or clarifying an existing agreement. The phase closes when the parties are satisfied and or the execution of the agreement is finished. Participants in this phase may include any or all of the actors previously involved or new specialists.

Example: The Ichthys offshore LNG project in Western Australia now approaching production in 2017, has spurred buyer and supplier negotiations in addition to those that created the project between discovery in 2006 and project start in 2012 [61].

4   Conclusion

This article asked whether evidence for macro phases from inception to completion of a negotiation could be found and whether the boundaries of those phases could be determined. Evidence is found in a variety of sources including current negotiation models, media reports about negotiations, and in transcripts previously published in academic sources. The Table 2 shows the phases with examples of their characteristics, content, and boundary definitions.

Previous negotiation models have not delivered a full overview of the negotiation process and have not clarified the macro phases through which a negotiation develops. The new model's contributions improve upon the existing models by providing an overall process that identifies the macro phases clearly with defensibly segmented phase boundaries and characteristics. Further, this model provides clear information about going forward or backward after each phase. It is hoped that the proposed model will spur improvement of this model and development of other models that reflect realities as well as ideal processes in organizations of all sorts.

Limitations

The author acknowledges various limitations including that the processes are dynamic and that their changing nature is difficult to model [63]. In particular this is true of the impact of information discovery as it emerges during the Interaction phase.

Table 2. Phase boundaries, outputs, and activities

	Start boundary	End boundary	Outputs and artifacts	Character	Main activities
Value network fit	Idea(s) mooted	Mandate to team; allies coopted	Board level statement; Written directive; Verbal statement; Meeting notes; Budget allocation	Strategic planning; Discovery	Strategic review; Consideration of alliances and high level goals
Deal design	Kickoff meeting with team	Start of main interaction with counterparties	Formal or informal plans; Meeting notes; Models; Spreadsheets;	Design of offers and deal packages; Discovery	Planning how to accomplish goals and present or react to other parties; Research
Interaction	Main interaction with counterparties	Agreement/ resolution	Written agreements; MOU; Overall agreement; Schedules of resources etc.	Iterative verbal and written interaction among parties in synchronous meetings or asynchronous media exchanges	Communicating ; proposing; reacting; joint problem solving; conceding; Building relationship;
Ratification	Presentation of agreement to final decision makers	Ratification; Signing ceremony; Disbursement of resources	Formal agreement; Contract; Ratification or certification; MOU	Completion and ratification	Presentation to/ agreement with final decision makers; Legal review
Evaluation and Monitoring	Review and evaluation; Monitoring of execution	End of evaluation or monitoring period	Evaluation and monitoring reports; Recommendations for improvement	Evaluation and monitoring	Evaluation of outcomes; Review of negotiators; Monitoring of execution
Follow up	Resolution of outstanding issues; Problem(s) arise	Completion of overall or subordinate agreements	Agreements; Addenda; Revisions; Reports; Project statements	Subordinate agreements and refinements	Completion of subordinate agreements; adjustment and refining of issues

References

1. Walton, R.E., McKersie, R.B.: A Behavioral Theory of Labor Negotiations: An Analysis of a Social Interaction System. ILR Press, Ithaca (1965)
2. Lewicki, R.J., Hiam, A., Olander, K.W.: Think Before You Speak: A Complete Guide to Strategic Negotiation. Wiley, Hoboken (1996)
3. Brooks, E.: Managing by Negotiations. Van Nostrand Rheinhold, New York (1984)
4. Spector, B.I.: Negotiation as a psychological process. J. Conflict Resolut. **21**, 607–618 (1977)
5. Bazerman, M.H., Curhan, J.R., Moore, D.A., Valley, K.L.: Negotiation. Ann. Rev. Psychol. **51**, 279–314 (2000)
6. Adair, W.L., Taylor, M., Chu, J., Ethier, N., Xiong, T., Okumura, T., Brett, J.M.: Effective influence in negotiation. Int. Stud. Manag. Organ. **43**, 6–25 (2013)
7. Craver, C.B.: Effective Legal Negotiation and Settlement. LexisNexis, New York (2012)
8. Zartman, I.W., Berman, M.R.: The Practical Negotiator. Yale University Press, New Haven (1983)
9. Lax, D.A., Sebenius, J.K.: 3D Negotiation: Powerful Tools to Change the Game in Your Most Important Deals. Harvard Business School Publishing, Boston (2006)
10. Ott, U.F., Prowse, P., Fells, R., Rogers, H.: The DNA of negotiations as a set theoretic concept: a theoretical and empirical analysis. J. Bus. Res. **69**, 3561–3571 (2016)
11. Colosi, T.: A core model of negotiation. In: Lewicki, R.J., Saunders, D.M., Minton, J.W., Barry, B. (eds.) Negotiation: Readings, Exercises, and Cases, 4th edn, pp. 292–297. McGraw-Hill Higher Education, Boston (2003)
12. English, T.: Negotiation as tension management: a model for business and other international transactions, Adelaide (2003)
13. Filzmoser, M., Vetschera, R.: A classification of bargaining steps and their impact on negotiation outcomes. Gr. Decis. Negot. **17**, 421–443 (2008)
14. Weingart, L.R., Olekalns, M.: Communication processes in negotiation: frequencies, sequences, and phases. In: Gelfand, M., Brett, J.M. (eds.) The Handbook of Negotiation and Culture, pp. 143–157. Stanford Business Books, Stanford (2004)
15. Holmes, M.E.: Phase structures in negotiation. In: Putnam, L.L., Roloff, M.E. (eds.) Communication and Negotiation, pp. 83–108. SAGE Publications, Thousand Oaks (1992)
16. Vetschera, R.: Negotiation processes: an integrated perspective. EURO J. Decis. Process. **1**, 135–164 (2013)
17. Filzmoser, M., Hippmann, P., Vetschera, R.: Analyzing the multiple dimensions of negotiation processes. Gr. Decis. Negot. **25**, 1169–1188 (2016)
18. Abbott, T.E.: Time phase model for hostage negotiation. Police Chief **6**, 169–186 (1986)
19. Brett, J., Weingart, L.R., Olekalns, M.: Baubles, bangles, and beads: modeling the evolution of negotiating groups over time. In: Blount, S., Mannix, B., Neale, M. (eds.) Research on Managing Groups and Teams, vol. 6, pp. 39–64. JAI Press, Greenwich (2003)
20. Fells, R., Rogers, H., Prowse, P., Ott, U.F.: Unraveling business negotiations using practitioner data. Negot. Confl. Manag. Res. **8**, 119–136 (2015)
21. Druckman, D., Olekalns, M.: Punctuated negotiations: transitions, interruptions, and turning points. In: Olekalns, M., Adair, W.L. (eds.) Handbook of Research on Negotiation, pp. 332–356. Edward Elgar Publishing Incorporated, Cheltenham (2013)
22. Putnam, L.L., Fuller, R.P.: Turning points and negotiation: the case of the 2007–2008 writers' strike. Negot. Confl. Manag. Res. **7**, 188–212 (2014)
23. Olekalns, M., Brett, J.M., Weingart, L.R.: Phases, transitions and interruptions: modeling processes in multi-party negotiations. Int. J. Confl. **14**, 191–211 (2003)

24. Koeszegi, S.T., Pesendorfer, E.M., Vetschera, R.: Data-driven phase analysis of e-negotiations: an exemplary study of synchronous and asynchronous negotiations. Gr. Decis. Negot. **20**, 385–410 (2011)
25. Weske, M.: Business Process Management: Concepts, Languages, Architectures. Springer, Heidelberg (2012)
26. Clegg, B.T., Boardman, J.T.: Process integration and improvement using systemic diagrams and a human-centred approach. Concurr. Eng. Res. Appl. **4**, 119–136 (1996)
27. Jeong, H.W.: International Negotiation: Process and Strategies. Cambridge University Press, Cambridge (2016)
28. Adair, W.L., Brett, J.M.: Culture and negotiation processes. In: Gelfand, M.J., Brett, J.M. (eds.) The Handbook of Negotiation and Culture, pp. 158–176. Stanford Business Books, Stanford (2004)
29. Douglas, A.: Industrial peacemaking. Columbia University Press, New York (1962)
30. Morley, I.E., Stephenson, G.M.: The Social Psychology of Bargaining. G. Allen & Unwin, London (1977)
31. Gulliver, P.H.: Disputes & Negotiations: A Cross-Cultural Perspective. Academic Press, Cambridge (1979)
32. Brooks, E., Odiorne, G.S.: Managing by Negotiations. Krieger, Malabar (1992)
33. Craver, C.B.: Effective Legal Negotiation and Settlement. Michie, Charlottesville (1986)
34. Heller, F.A., Drenth, P., Koopman, P., Rus, V.: Decisions in Organizations: A Three-Country Comparative Study. Sage, London (1988)
35. Boehm, B., Bose, P., Horowitz, E., Lee, M.J.: Software requirements negotiation and renegotiation aids: a theory-w based spiral approach 1. In: Proceedings of ICSEE, vol. 95, pp. 1–18 (1997)
36. Straus, D.: Designing a consensus building process using a graphic road map. In: Susskind, L., McKearnen, S., Thomas-Lamar, J. (eds.) The Consensus Building Handbook, pp. 137–168. SAGE Publications, Thousand Oaks (1999)
37. Lopes, F., Mamede, N., Novais, A.Q., Coelho, H.: Towards a generic negotiation model for intentional agents. In: Proceedings of 11th International Work Database and Expert System Applications (2000)
38. In, H., Olson, D., Rodgers, T.: A requirements negotiation model based on multi-criteria analysis. In: Proceedings Fifth IEEE International Symposium on Requirements Engineering, Toronto, ON, pp. 312–313. IEEE (2001)
39. Adair, W.L., Brett, J.M.: The negotiation dance: time, culture, and behavioral sequences in negotiation. Organ. Sci. **16**, 33–51 (2005)
40. Demirkan, H., Goul, M., Soper, D.S.: Service level agreement negotiation: a theory-based exploratory study as a starting point for identifying negotiation support system requirements. In: Proceedings of the Annual Hawaii International Conference on System Sciences, p. 37b (2005)
41. Mutual gains approach. http://www.cbuilding.org
42. Lewicki, R.J., Hiam, A.: Mastering Business Negotiation: A Working Guide to Making Deals and Resolving Conflict. Jossey-Bass, San Francisco (2010)
43. Taylor, P.J.: A cylindrical model of communication behaviour in crisis negotiations. Hum. Commun. Res. **28**, 7–48 (2002)
44. Turan, N., Dai, T., Sycara, K., Weingart, L.R.: Toward a unified negotiation framework: leveraging strengths in behavioral and computational communities. In: Sycara, K., Gelfand, M., Abbe, A. (eds.) Advances in Group Decision and Negotiation: Models for Intercultural Collaboration and Negotiation, pp. 53–65. Springer, Netherlands (2013)

45. Phalp, K.T.: The CAP framework for business process modelling. Inf. Softw. Technol. **40**, 731–744 (1998)
46. Ting-Toomey, S.: The matrix of face: an updated face-negotiation theory. In: Gudykunst, W.B. (ed.) Theorizing about Intercultural Communication, pp. 71–92. Sage, Thousand Oaks (2005)
47. Graham, J.L., Lam, N.M.: The Chinese negotiation. Harvard Bus. Rev. **81**, 82–91 (2003)
48. Project Management Institute: A Guide to the Project Management Body of Knowledge. Project Management Institute, Newtown Square (2015)
49. Binder, J.: Global Project Management. Gower, Farnham (2007)
50. Mascitelli, R.: The Lean Product Development Guidebook: Everything Your Design Team Needs to Improve Efficiency and Slash Time-To-Market: Technology Perspectives. Northridge, CA (2007)
51. Thamhain, H.J.: Management of Technology: Managing Effectively in Technology-Intensive Organizations. Wiley, Hoboken (2013)
52. Allee, V.: Reconfiguring the value network. J. Bus. Strategy. **21**, 36–39 (2000)
53. Minchin, T.J.: Shutting down "Big Brown": reassessing the 1997 UPS strike and the fate of American labor. Labor Hist. **53**, 541–560 (2012)
54. Miller, B.K.: Issues management: the link between organization reality and public perception. Publ. Relations Q. **44**, 5–11 (1997)
55. Gupta, P., Damouni, N., Sandle, P.: How a desperate HP suspended disbelief for autonomy deal. http://www.reuters.com/article/2012/11/30/us-hp-autonomy-idUSBRE8AT09X20121130
56. Shapiro, D.: The HPQ/AU scandal: what went wrong? J. Corp. Account. Financ. **24**, 49–53 (2013)
57. Moberg, D.: The UPS Strike: lessons for labor. Work. USA **1**, 11–29 (1997)
58. Vuorela, T.: How does a sales team reach goals in intercultural business negotiations? Case Study. English Specif. Purp. **24**, 65–92 (2005)
59. Vandaele, D., Rangarajan, D., Gemmel, P., Lievens, A.: How to govern business services exchanges: contractual and relational issues. Int. J. Manag. Rev. **9**, 237–258 (2007)
60. Raiffa, H.: Post-settlement settlements. Negot. J. **1**, 9–12 (1985)
61. INPEX: The Ichthys LNG project, Perth (2016)
62. Baber, W.W.: Looking inside Japanese-Japanese intracultural business negotiation. Kyoto Econ. Rev. **85**, 104–134 (2016)
63. Lindsay, A., Downs, D., Lunn, K.: Business processes - attempts to find a definition. Inf. Softw. Technol. **45**, 1015–1019 (2003)

Decision Problems in Requirements Negotiations – Identifying the Underlying Structures

Annika Lenz[✉] [iD] and Mareike Schoop

Department of Information Systems I, University of Hohenheim, Stuttgart, Germany
{annika.lenz,schoop}@uni-hohenheim.de

Abstract. In the preparation phase of negotiation processes, the decision problem needs to be identified to assign preferences and thus enable offer evaluation. However, in software requirements negotiations, identification of the required decision problem structure is not easy, since requirements negotiations vary due to organisational factors such as stakeholders involved or the software development method applied. This paper identifies decision problem structures in software requirements negotiations using a literature-based research approach. In doing so, a matrix of decision relevant information in software requirements negotiations and their representation in a negotiation context is developed. The matrix can be utilised as a framework to select appropriate scenarios of decision problem structures in software requirements negotiations.

Keywords: Decision problem structure · Software requirements negotiation · Negotiation problem · Decision support · MCDM

1 Introduction

Requirements negotiations are iterative processes, in which involved parties with (usually conflicting) individual goals jointly seek to reach an overall goal, namely agreeing on a software development process and outcome [1]. One of their main characteristics is that involved parties need to make decisions during these processes [1, 2]. Due to domain-related characteristics of requirements negotiations, they differ in terms of the role of the negotiating parties, the type of negotiation issues, and thus the decisions to be made. That is because the context of requirements negotiations depends on software project related factors. The negotiation parties in a software organisation include, on the one hand, individual people such as product manager, project manager, or architect, and, on the other hand, groups of people who represent negotiation parties such as customer, supplier, management of a company, or development team. Every stakeholder brings individual and role-specific knowledge, capabilities, and skills [3] as well as role-dependent negotiation issues and decisions to be made during the process of requirements negotiations.

Moreover, the software project guidelines determining the software development method influence the scope and frequency of requirements negotiations. In traditional software development, requirements negotiations at the beginning of the project are essential. Requirements engineering mainly takes place at the project start defining the

© Springer International Publishing AG 2017
M. Schoop and D.M. Kilgour (Eds.): GDN 2017, LNBIP 293, pp. 120–131, 2017.
DOI: 10.1007/978-3-319-63546-0_9

scope of the software project. However, the advent of agile software development (which aims at delivering software more speedily and frequently [4]) saw software development teams deliver functional software in shorter development cycles. Hence, agile software development as an iterative incremental development approach requires more frequent requirements negotiations of smaller scope based on user requirements and requirements changes [4, 5].

Dedicated electronic support provides great advantages for communication and decision making in negotiations [6]. To support decisions within these various processes of requirements negotiations, is our ultimate goal. Negotiation research provides approaches for decision support (for an overview of methods and systems see [7]). However, standard methods used to support negotiations are not straightforwardly suitable to support requirements negotiations in their present form. The decision process in requirements negotiations is a different one facing incomplete, missing and changing information throughout the process. Thus adapted solutions are required [1].

Requirements negotiations can be divided into the following three phases: preparation, negotiation, and settlement [8, 9]. In the pre-negotiation phase – whether electronically supported or not –, a vision of the decision problem needs to be developed in detail [10]. The definition of the decision problem and thus precisely structuring of the decision problem is of utmost importance, since the structure may impact the negotiation process and outcome [11]. In alignment with the aspect of problem-orientation of negotiation support systems (NSSs), which evolved from decision support systems, the user must be helped to understand the problem structure [12]. Moreover, only if the decision problem, which needs to be optimised [10], is defined, electronic decision support can be provided [13].

To pave the way for future work on this topic, we bring the task of identifying the decision problem as one of the first, indispensable tasks in each (requirements) negotiation process into focus. Consequently, our research question is: *How is the decision problem in requirements negotiations structured?*

The overall aim is to enable decision support for one actor from an individual perspective. In this paper, the decision problem structure is investigated independently of the type of negotiation (bilateral or multilateral). In doing so, our paper firstly contributes to the preparation phase in requirements negotiations – independently if electronically supported or not – and secondly to the enablement of decision support for requirements negotiations.

To this end, we describe the structure of decision problems in the context of decision theory as well as the decision problem structure, which is supported by NSSs. Subsequently, domain information which is relevant for the decision problem is identified and categorised. The resulting categories are then transferred to the decision problem context to extract the decision problem structure in requirements negotiations. A literature-based research approach has been chosen to accomplish this aim.

The remainder of this paper is structured as follows. In Sect. 2, we outline the theoretical background regarding decision problems. In Sect. 3, information relevant to the decision problem in requirements negotiations is described and applied to a decision problem structure following the two paradigms of software development, namely traditional requirements engineering and agile requirements engineering, resulting in a matrix to identify decision relevant information in a negotiation context. In Sect. 4, our findings are briefly discussed.

2   Decision Problem Structure in Negotiations

A multi-criteria decision problem consists of the objective, optionally lower-level objectives, attributes, and attribute values [14]. The objective of a decision problem indicates the direction to strive for. An objective can be subdivided into lower-level objectives of more detail, also called sub-objectives. The lowest-level objectives may be associated with attributes, which will indicate the degree of meeting the objective. Attributes are used to measure the objective according to their attribute values. The decision problem itself is organised in a hierarchy with infinite optional levels [14, 15].

In negotiation analysis theory, a negotiation template specifies the structure of the decision problem in detail [10]. The negotiation template design comprises the negotiation issues and all feasible options to resolve these issues. Thus, in a negotiation context, the objective is associated with attributes, which are negotiable and to be negotiated. Such attributes are called negotiation issues. For each issue, possible resolutions are assigned, to which we refer as negotiation alternatives [16]. We treat alternatives as negotiable values. Hence, a decision problem in negotiations is in general structured by an objective, negotiation issues, and negotiation alternatives, see Fig. 1.

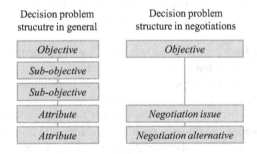

Fig. 1. Decision problem structure in general and in negotiations.

State-of-the-art NSSs such as Inspire [17], Negoisst [6, 18], or SmartSettle [19] support this basic decision problem structure. Such systems follow multi-attribute utility theory (MAUT), predominantly utilising linear additive utility functions, which are calculated based on an individual's preferences [14]. The objective of the decision problem is to maximise the utility of a negotiation offer. A negotiation offer's utility is the sum of the partial utilities of the selected negotiation alternatives based on the corresponding negotiation issue's weight. The general decision problem structure in negotiations in conjunction with the concept of linear additive utility functions is applied in current NSSs for negotiation analysis.

3   Requirements Negotiations

In the following, we will investigate problem structures in requirements negotiations following two paradigms of software development methods, namely *the traditional software development method*, such as the waterfall method, abbreviated as traditional requirements negotiations (Sect. 3.1), as well as the *agile software development method*, such as Scrum, abbreviated as agile requirements negotiations (Sect. 3.2). In Sect. 3.3 we present their implication in the form of a matrix.

3.1   Decision Problem Structures in Traditional Requirements Negotiations

The aim of requirements negotiations is to get an agreed-upon sound set of requirements [1]. Through requirements negotiations, stakeholders make trade-offs between the desired system functionality, the technology to be applied, the project schedule as well as project cost [20]. Thus, not only required system functions but also non-functional requirements, the technology stack, and conditions are to be negotiated [21].

Packages. Packages are used for structuring reasons. Structuring may be based on functions, e.g. when requirements are bundled into features, since a software project covers a huge number of requirements.

Firstly, packages are used to assign negotiable requirements to sub-negotiations, which are conducted separately. In NSSs, a hierarchy level for structuring is not yet applied. However, the objective of decision problems can be divided into sub-objectives. Therefore, we expand the negotiation context by a sub-objective level to cover *packages as sub-objectives*.

Secondly, packages describe negotiation issues themselves, e.g. in negotiations about whether to implement features at all or when to implement features. So, we map *packages also to negotiation issues*.

Requirements. A requirement is "(1) A condition or capability needed by a user to solve a problem or achieve an objective. (2) A condition or capability that must be met or possessed by a system or system component to satisfy a contract, standard, specification, or other formally imposed documents. (3) A documented representation of a condition or capability as in (1) or (2)" [22, p. 65]. This definition covers different types of requirements, namely functional requirements, quality requirements, and constraints [23]. Since requirements negotiation is about negotiating requirements, *requirements are treated as negotiation issues*.

However, if the negotiation takes place on a higher level, in which packages of requirements are negotiated, the requirements themselves can be utilised to describe these packages, in which case the *requirements represent attributes*. Since these attributes are neither negotiation issues nor negotiation alternatives, but are used for description as attributes in general decision problem structures, we keep the term attribute. Since decision problems in negotiations do not yet cover such kind of attributes, the negotiation context is also expanded by attributes for negotiation issues.

Solutions. Solutions are implementation scenarios for requirements. While requirements describe the problem to be addressed in the software development process and thus specify "what" is to be developed, solutions describe the solution space to this problem, and thus specify "how" it is to be developed [23].

Therefore, considering requirements as negotiation issues, *solutions describe negotiation alternatives* [21]. Solutions may cover a single requirement or a whole package. The latter is the case, e.g. when choosing technologies to be deployed, which hold for a whole package [24]. Hence, if packages are formulated as negotiation issues, *solutions address negotiation alternatives* as well.

Criteria. In requirements negotiations, requirements are assessed in preparation for decision making. This has the benefit that the decision makers can assign their preferences based on the assessment rather than assigning their preferences directly to the requirements. In some approaches providing decision support, requirements are assessed with respect to their business value and development effort [25, 26]. Other approaches leave the criteria definition to the project stakeholders. In this case, they generate relevant criteria at the beginning of the negotiation, e.g. by an initial brainstorming [27]. Moreover, a two-level approach prioritises features with respect to business goals of the organisation and utilises ease of realisation and business value to prioritise low-level requirements [28]. Thakurta (2016) identifies a plethora of thirteen requirements attributes, which influence requirements prioritisation [29]. Thus, *criteria address requirements as attributes for negotiation issues*.

However, criteria may also be used to assess packages directly. Likewise, criteria for packages facilitate the decision makers to decide on a better valuation basis. Compared with criteria for requirements, criteria for packages have the advantage that information does not need to be elicited in such detail. So, *criteria for packages also represent attributes for negotiation issues*.

Criteria may also be applied to solutions for assessment reasons. Thus, criteria may also be used to *describe solutions* and hence used as *attributes to describe negotiation alternatives*.

Criteria are also used to define overall project constraints or contract conditions, such as project duration or project budget. In this case, criteria are negotiable themselves and thus, formulated as *negotiation issues* among requirements and solutions [21].

Possible Decision Problem Structures. From above, different decision problem structures following traditional software development methods can be derived, see Fig. 2. We observe semantic differences in the negotiation context but also differences in the hierarchical structure. NSSs allow only a two-level structure of negotiation issues and alternatives (e.g. [6, 18]). However, the number of levels in requirements negotiations is potentially higher than in usual negotiations due to domain characteristics. Thus, current NSSs do not completely facilitate decision support for requirements negotiations.

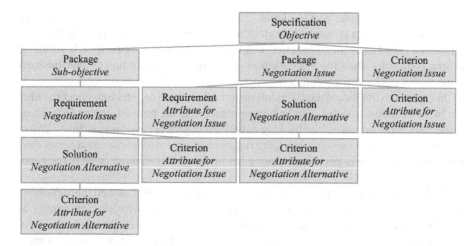

Fig. 2. Possible decision problem structures in traditional requirements negotiations.

3.2 Decision Problem Structures in Agile Requirements Negotiations

Agile software development claims to deliver working software early and more frequently as well as collaborate closely with the customer to satisfy the customer [30]. Fulfilling these aims, the requirements engineering process needs to be adapted to accomplish its goals [31]. The difference of the requirements engineering process in traditional software methods and in agile software methods, however, is not that it would include different activities, but that in agile requirements engineering the requirements engineering activities take place iteratively in each development cycle, namely requirements elicitation, documentation, and validation as well as negotiation, and are not sequential but intermingled and addressed together [31].

Nonetheless, in the pre-negotiation phase, the decision problem, which is to be negotiated, must be identified. This is done in a more extensive manner for the first negotiation, while in each iteration, the decision problem may be derived from already identified negotiation issues (from the product backlog or the sprint backlog) or from changed negotiation issues.

In the requirements analysis and negotiation activities, the focus is on refining, changing, and prioritising requirements [31]. These activities are performed during assembling the backlog, which contains a list of items, which are desired to be implemented [32]. The backlog items cover requirements, which are in general specified in user stories [32, 33]. These requirements are argued to be negotiable [32]. From the product backlog, the release backlog is derived, which covers a fragment of the required functionality [32]. Based on the release backlog, the sprint backlog defines the requirements, which are to be implemented in the next sprint [32]. Hence, release and sprint planning are main activities, in which requirements bundles are negotiated, which are to be implemented in the next release, respectively in the next sprint.

Some studies claim that the product owner is responsible for prioritisation of the backlog [e.g. 34], however others argue that (in the sprint planning), product managers and developers negotiate backlog items [35]. Especially, if different customer groups are involved, who are concerned about different required functionality, negotiation is performed [31]. Hence, main negotiation activities in each development cycle comprise to seek consensus about the *prioritisation of requirements* according to their business value or risk [33] and accordingly their *assignment* to iterations.

In contrast to traditional requirements negotiations, these negotiations concern rather the "what" dimension (what is to be implemented) than the how dimension [33]. Thus, in the case of prioritisation, the backlog covers the objective *(product backlog – objective; sprint backlog – sub-objective)*, the *requirements the negotiation issues*, and *their prioritisation*, which is negotiated among the involved stakeholders, covers the *negotiation alternatives*.

In the assignment negotiation, the respective backlog covers as well the *sub-objective*, and the *requirements the negotiation issues*. Here, the *prioritisation applies* as criteria for the requirements and thus, *attributes for negotiation issues*, while the *iteration covers the negotiation alternatives* in terms of to which iteration a requirement is assigned (or if it is assigned to the next iteration, or if at all). Figure 3 shows possible decision problem structures in agile requirements negotiations.

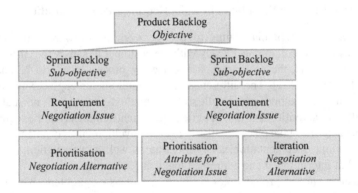

Fig. 3. Possible decision problem structures in agile requirements negotiations.

Applied to the terminology identified in Sect. 3.1, the product backlog represents the specification, the release, and sprint backlog represent a package, the prioritisation represents a criterion, and the iteration represents a solution.

3.3   Matrix of Possible Decision Problem Structures

In Sect. 2, we found out that a decision problem in general may consist of the objective, sub-objectives, and attributes, whereas the decision problem in negotiations consists of an objective, negotiation issues, and negotiation alternatives. From the above described decision-relevant information in the domain context, which is categorised into packages, requirements, solutions, and criteria, we develop the following matrix to transfer these categories into a negotiation context, see Table 1.

Table 1. Matrix of possible decision problem structures in requirements negotiations. (NP: General decision problem structure in negotiations; X: Applies; T: In traditional requirements negotiations; A: In agile requirements negotiations).

	NP		NP	NP		
	Objective	Sub-objective	Negotiation issue	Negotiation alternative	Attribute for negotiation Issue	Attribute for negotiation alternative
Specification	X (T, A)					
Package		X (T, A)	X (T)			
Requirement			X (T, A)		X (T)	
Solution				X (T, A)		
Criterion			X (T)	X (A)	X (T, A)	X (T)

This matrix allows to determine relevant decision problem structures in requirements negotiations and shows that in certain scenarios, the negotiation context needs to be expanded by a structuring level and/or by a level, which provides more information.

Regarding this diversity, there is no unique approach to apply a decision problem structure to the decision problem to be solved. From a customer's perspective, one could base the decision on a set of requirements on cost and time. Other decision makers may prefer to base their decision on how the requirements are implemented (i.e. the architectural design) rather than cost and time. A developer's perspective might rather include available resources. Consequently, there is more than one way to design and structure the decision problem.

4   Discussion and Conclusion

We have shown that there is no uniform overall decision problem structure for requirements negotiations. The developed possible decision problem structures in requirements negotiations and their resulting matrix facilitates to accomplish the identification of the decision problem of the use case at hand during the pre-negotiation phase. This task is of prime importance, independently if requirements negotiations are carried out with electronic support or without [13]. In requirements negotiation, where teams are dispersed, electronic support of the pre-negotiation phase is of great benefit. Where this task is performed face to face, which is often the case in agile requirements negotiation [31, 33], the advantages of electronic media can be exploited using synchronous tools (e.g. like the EasyWinWin approach [36]). Moreover, synergies can be enabled by providing interfaces to requirements management and software development tools to complement requirements negotiation support systems. Furthermore, it is suggested not to support requirements negotiations electronically using a single medium, but support them by a mix of media [37].

We chose a literature-based approach. Thus, one limitation of our approach is that our results reflect only those requirements negotiations concepts and approaches that have been published. However, the literature utilised builds inter alia upon empirical results. Therefore, this limitation is negligible.

Current requirements negotiation systems providing decision support can be grouped into automated approaches, recommender systems, and interactive approaches (for an overview see [38]). Automated approaches negotiate autonomously based on criteria specified prior to the negotiation process. Recommender systems provide recommendations for group decision making during requirements negotiations, e.g. [39]. Interactive decision support in requirements negotiations utilises an analytic hierarchy process, e.g. [25], or simple additive weighting approaches, e.g. [27] (variants of the Easy-WinWin approach). In this group, some systems (such as [25, 26]) require the fixed criteria business value and ease of realisation, whilst other (such as [27]) provide the opportunity to define individual criteria. Thus, the support level of these three groups is a different one. However, what all approaches have in common is that they assume a certain structure of the decision problem. EasyWinWin approaches apply requirements (i.e. win conditions), solutions (i.e. resolution options), and criteria. A two-level approach, based on EasyWinWin and utilising TOPSIS, includes packages [40]. Each system supports one and only one decision problem structure, which is presumed for each negotiation. Negotiation of criteria in terms of contract conditions such as the price is not considered in these approaches. They focus on the negotiation of requirements.

Requirements negotiation systems which do not presume a specific decision problem structure focus on supporting group collaboration or enhancing participation, e.g. [41–43].

The variability of the decision problem structure makes it difficult to provide decision support in requirements negotiations. The more flexible the decision support approach, i.e. the more scenarios it can support, the more complex both development and usage of such a system are. From this point of view, it is to consider, which scenarios to concentrate on, e.g. based on their importance or frequency. To develop a system, which supports reasonable decision problems in traditional and agile requirements negotiations, is our ultimate goal and the present paper contributes to its conceptual design.

Acknowledgements. The authors wish to acknowledge the valuable comments and suggestions made by the anonymous reviewers and gratefully acknowledge the funding provided by the Faculty of Business, Economics, and Social Sciences at the University of Hohenheim within the research area "Negotiation Research – Transformation, Technology, Media, and Costs".

References

1. Lenz, A., Schoop, M., Herzwurm, G.: Requirements analysis as a negotiation process. In: Kaminski, B., Kersten, G., Szufel, P., Jakubczyk, M., Wachowicz, T. (eds.) The 15th International Conference on Group Decision & Negotiation Letters, pp. 303–309. Warsaw School of Economics Press (2015)
2. Aurum, A., Wohlin, C.: The fundamental nature of requirements engineering activities as a decision-making process. Inf. Softw. Technol. **45**, 945–954 (2003). doi:10.1016/S0950-5849(03)00096-X

3. Fricker, S., Grünbacher, P.: Negotiation constellations – method selection framework for requirements negotiation. In: Paech, B., Rolland, C. (eds.) REFSQ 2008. LNCS, vol. 5025, pp. 37–51. Springer, Heidelberg (2008). doi:10.1007/978-3-540-69062-7_4

4. Beck, K., Beedle, M., van Bennekum, A., Cockburn, A., Cunningham, W., Fowler, M., Grenning, J., Highsmith, J., Hunt, A., Jeffries, R., et al.: Manifesto for agile software development. http://agilemanifesto.org/

5. AL-Ta'ani, R.H., Razali, R.: Prioritizing requirements in agile development: a conceptual framework. Procedia Technol. **11**, 733–739 (2013). doi:10.1016/j.protcy.2013.12.252

6. Schoop, M.: Support of complex electronic negotiations. In: Kilgour, D.M., Eden, C. (eds.) Handbook of Group Decision and Negotiation, vol. 4, pp. 409–423. Springer, Netherlands (2010). doi:10.1007/978-90-481-9097-3_24

7. Reiser, A.: Entscheidungsunterstützung in elektronischen Verhandlungen. Eine Analyse unter besonderer Berücksichtigung von unvollständigen Informationen. Springer Fachmedien Wiesbaden. Springer Gabler, Wiesbaden (2013)

8. Reiser, A., Krams, B., Schoop, M.: Requirements negotiation in consideration of dynamics and interactivity. In: Svensson, R.B., Berry, D., Daneva, M., Dörr, J., Fricker, S.A., Herrmann, A., Herzwurm, G., Kauppinen, M., Madhavji, N.H., Mahaux, M., Paech, B., Penzenstadler, B., Pietsch, W., Salinesi, C., Schneider, K., Seyff, N., van de Weerd, I., (eds.) Proceedings of the Workshops RE4SuSy, REEW, CreaRE, RePriCo, IWSPM and the Conference Related Empirical Study, Empirical Fair and Doctoral Symposium 18th International Working Conference on Requirements Engineering: Foundation for Software Quality, ICB Research Report No. 52, pp. 163–174. Universität Duisburg-Essen (2012). ISSN:1860-2770 (Print), ISSN:1866-5101 (Online)

9. Grünbacher, P., Köszegi, S., Biffl, S.: Stakeholder value proposition elicitation and reconsiliation. In: Biffl, S., Aurum, A., Boehm, B., Erdogmus, H., Grünbacher, P. (eds.) Value Based Software Engineering, pp. 133–154. Springer, Heidelberg (2006). doi: 10.1007/3-540-29263-2_7

10. Górecka, D., Roszkowska, E., Wachowicz, T.: The MARS approach in the verbal and holistic evaluation of the negotiation template. Group Decis. Negot. **25**, 1097–1136 (2016). doi: 10.1007/s10726-016-9475-9

11. Mumpower, J.L.: The judgment policies of negotiators and the structure of negotiation problems. Manage. Sci. **37**, 1304–1324 (1991). doi:10.1287/mnsc.37.10.1304

12. Kersten, G., Lai, H.: Electronic negotiations: foundations, systems, and processes. In: Kilgour, D.M., Eden, C. (eds.) Handbook of Group Decision and Negotiation, vol. 4, pp. 361–392. Springer, Dordrecht (2010). doi:10.1007/978-90-481-9097-3_22

13. Simons, T., Tripp, T.M.: The negotiation checklist. In: Lewicki, R.J., Saunders, D.M., Barry, B. (eds.) Negotiation: Readings, Exercises, and Cases, pp. 34–47. McGraw-Hill Higher Education, New York (2010)

14. Keeney, R.L., Raiffa, H.: Decisions with Multiple Objectives: Preferences and Value Tradeoffs. Wiley, New York (1976)

15. Danielson, M., Ekenberg, L.: The CAR method for using preference strength in multi-criteria decision making. Group Decis. Negot. **25**, 775–797 (2016). doi:10.1007/s10726-015-9460-8

16. Raiffa, H., Richardson, J., Metcalfe, D.: Negotiation Analysis: The Science and Art of Collaborative Decision Making. Belknap Press of Harvard University Press, Cambridge (2002)

17. Kersten, G.E., Noronha, S.J.: WWW-based negotiation support: design, implementation, and use. Decis. Support Syst. **25**, 135–154 (1999). doi:10.1016/S0167-9236(99)00012-3

18. Schoop, M., Jertila, A., List, T.: Negoisst: a negotiation support system for electronic business-to-business negotiations in e-commerce. Data Knowl. Eng. **47**, 371–401 (2003). doi:10.1016/S0169-023X(03)00065-X

19. Thiessen, E.M., Soberg, A.: SmartSettle described with the Montreal taxonomy. Group Decis. Negot. **12**, 165–170 (2003). doi:10.1023/A:1023025106197

20. Grünbacher, P., Seyff, N.: Requirements negotiation. In: Aurum, A., Wohlin, C. (eds.) Engineering and Managing Software Requirements, pp. 143–162. Springer, Heidelberg (2005). doi:10.1007/3-540-28244-0_7

21. Herzwurm, G., Schoop, M., Reiser, A., Krams, B.: E-requirements negotiation: electronic negotiations in the distributed software development. In: Mattfeld, D.C., Robra-Bissantz, S. (eds.) Multikonferenz Wirtschaftsinformatik 2012. Tagungsband der MKWI 2012, pp. 1859–1870. Gito; Univ.-Bibl, Berlin, Braunschweig (2012)

22. IEEE standard glossary of software engineering terminology. 610.12-1990. Institute of Electrical and Electronics Engineers, New York (1990)

23. Pohl, K.: Requirements Engineering: Fundamentals, Principles, and Techniques. Springer, Heidelberg (2010)

24. Franch, X., Carvallo, J.P.: A quality-model-based approach for describing and evaluating software packages. In: Proceedings, 9–13 September 2002, Essen, Germany, pp. 104–111. IEEE Computer Society, Los Alamitos (2002). doi:10.1109/ICRE.2002.1048512

25. Ruhe, G., Eberlein, A., Pfahl, D.: Quantitative WinWin – a new method for decision support in requirements negotiation. In: Proceedings of the 14th International Conference on Software Engineering and Knowledge Engineering (SEKE), pp. 159–166 (2002). doi:10.1145/568760.568789

26. Boehm, B., Grünbacher, P., Briggs, R.O.: Developing groupware for requirements negotiation: lessons learned. IEEE Softw. **18**, 46–55 (2001). doi:10.1109/52.922725

27. In, H., Olson, D.: Requirements negotiation using multi-criteria preference analysis. J. Comput. Sci. **10**, 306–325 (2004)

28. Kukreja, N., Boehm, B.: Integrating collaborative requirements negotiation and prioritization processes: a match made in heaven. In: International Conference on Software and System Process (ICSSP), pp. 141–145 (2013). doi:10.1145/2486046.2486071

29. Thakurta, R.: Understanding requirement prioritization artifacts: a systematic mapping study. Requirements Eng., 1–36 (2016). doi:10.1007/s00766-016-0253-7. First Online: 10 May 2016

30. Beck, K., Beedle, M., van Bennekum, A., Cockburn, A., Cunningham, W., Fowler, M.: Principles behind the Agile Manifesto. http://agilemanifesto.org/principles.html

31. Ramesh, B., Cao, L., Baskerville, R.: Agile requirements engineering practices and challenges: an empirical study. Inf. Syst. J. **20**, 449–480 (2010). doi:10.1111/j.1365-2575.2007.00259.x

32. Clarke, R.J., Kautz, K.: What's in a user story: IS development methods as communication. In: Proceedings of the 23rd International Conference on Information Systems Development, pp. 356–364. University of Zagreb, Croatia (2014).

33. Inayat, I., Salim, S.S., Marczak, S., Daneva, M., Shamshirband, S.: A systematic literature review on agile requirements engineering practices and challenges. Comput. Hum. Behav. **51**, 915–929 (2015). https://doi.org/10.1016/j.chb.2014.10.046

34. Moe, N.B., Aurum, A., Dybå, T.: Challenges of shared decision-making: a multiple case study of agile software development. Inf. Softw. Technol. **54**, 853–865 (2012). doi:10.1016/j.infsof.2011.11.006

35. Vlaanderen, K., Jansen, S., Brinkkemper, S., Jaspers, E.: The agile requirements refinery. Applying SCRUM principles to software product management. Inf. Softw. Technol. **53**, 58–70 (2011). doi:10.1016/j.infsof.2010.08.004

36. Boehm, B., Kitapci, H.: The WinWin approach: using a requirements negotiation tool for rationale capture and use. In: Dutoit, A.H., McCall, R., Mistrík, I., Paech, B. (eds.) Rationale Management in Software Engineering, pp. 173–190. Springer, Heidelberg (2006). doi: 10.1007/978-3-540-30998-7_8

37. Damian, D., Lanubile, F., Mallardo, T.: On the need for mixed media in distributed requirements negotiations. IIEEE Trans. Softw. Eng. **34**, 116–132 (2008). doi:10.1109/TSE. 2007.70758

38. Lenz, A., Schoop, M., Herzwurm, G.: Electronic requirements negotiation – a literature survey on the state-of-the-art. In: UK Academy for Information Systems Conference, Proceedings of UKAIS 2016 (Oxford), paper 23 (2016)

39. Felfernig, A., et al.: Group decision support for requirements negotiation. In: Hutchison, D., et al. (eds.) Advances in User Modeling, pp. 105–116. Springer, Heidelberg (2012). doi: 10.1007/978-3-642-28509-7_11

40. Kukreja, N., Payyavula, S.S., Boehm, B., Padmanabhuni, S.: Value-based requirements prioritization: usage experiences. Procedia Comput. Sci. **16**, 806–813 (2013). doi:10.1016/j.procs.2013.01.084

41. Walle, B., Campbell, C., Deek, F.P.: The impact of task structure and negotiation sequence on distributed requirements negotiation activity, conflict, and satisfaction. In: Krogstie, J., Opdahl, A., Sindre, G. (eds.) CAiSE 2007. LNCS, vol. 4495, pp. 381–394. Springer, Heidelberg (2007). doi:10.1007/978-3-540-72988-4_27

42. Kukreja, N.: Winbook: a social networking based framework for collaborative requirements elicitation and WinWin negotiations. In: 34th International Conference on Software Engineering (ICSE), pp. 1610–1612 (2012)

43. Renzel, D., Behrendt, M., Klamma, R., Jarke, M.: Requirements bazaar: social requirements engineering for community-driven innovation. In: 21st IEEE International Requirements Engineering Conference Proceedings, pp. 326–327 (2013). doi:10.1109/RE.2013.6636738

The Role of Sentiment and Cultural Differences in the Communication Process of e-Negotiations

Nil-Jana Akpinar[1], Simon Alfano[1(✉)], Gregory Kersten[2], and Bo Yu[2]

[1] University of Freiburg, Freiburg im Breisgau, Germany
niljana.akpinar@gmail.com, simon.alfano@is.uni-freiburg.de
[2] Concordia University Montréal, Montréal, Canada
{gregory.kersten,bo.yu}@concordia.ca

Abstract. Research shows that cultural differences affect negotiation processes and outcomes in many different ways. In this paper, we examine the interactions between communication processes, language, and cultural differences in dyadic e-negotiations. We use textual analysis methods to measure the language sentiment (also referred to as tone) of the messages. We make use of 9,703 messages (and offers with messages) in 1,147 negotiations conducted with the web-based negotiation support system Inspire. We find evidence that the more positive a message's sentiment, the more positive the sentiment of the next message. Our results indicate that this effect is less pronounced in intercultural negotiations. Furthermore, we observe higher payoffs for the party who initiates the conversation. Initiation reduces the risk of obtaining only a low payoff. Some cultural groups, such as German-speaking Western Europeans, emerge as particularly likely to initiate a negotiation.

Keywords: Communication process · Culture · Electronic negotiation · Initiation behavior · Sentiment analysis

1 Introduction

Negotiation is an important means of conflict resolution; it is *"a discussion between two or more parties with the apparent aim of resolving a divergence of interests"* [31, p. 2]. The investigation of negotiation processes yields insight into negotiation behavior and its communicational interdependencies [42]. Recent research on emotions in negotiations focuses on the strategic display of emotions and their perception by the negotiation partner [3,7,35]. Other research studies the antecedents of emotions, including the negotiators' cultural differences [2,22]. Cultural differences can lead to difficulties in communication, not least in the perception of emotions, influencing negotiation processes and outcomes. In order to understand these effects, the use of dictionaries comprising words associated with emotions coded in language has emerged as a method. Sole reliance on linguistic emotional cues is insufficient when the negotiators are able to employ non-verbal communication. The use of anonymous e-negotiations allows study of

© Springer International Publishing AG 2017
M. Schoop and D.M. Kilgour (Eds.): GDN 2017, LNBIP 293, pp. 132–144, 2017.
DOI: 10.1007/978-3-319-63546-0_10

emotions expressed in language, as well as the underlying cultural and personal preconceptions.

Hine et al. [16] show that agreeable and positive language impacts the likelihood of reaching an agreement. Analysis of anonymous, cross-cultural e-negotiations provides an opportunity to study the role of culture on emotions expressed in language.

The aim of the current study is to understand the interactions between language, culture, and negotiation processes and outcomes in dyadic e-negotiations. In particular, we analyze whether the language sentiment of a message influences the language sentiment of the subsequent message. We also analyze whether inter- and intracultural negotiations differ. Furthermore, we examine the influence of initiation on a negotiator's payoff and the impact of cultural factors on initiation.

To this end, the study makes uses of textual analysis methods in order to extract linguistic aspects of expressed emotions and to calculate the positivity or negativity of the negotiators' messages.

In the following, Sect. 2 reviews the literature and develops our hypotheses. Section 3 provides a description of our data and depicts our variables. Section 4 illustrates the results of our analyses. Interpretation of these results is given in Sect. 5.

2   Research Background and Hypotheses

2.1   Language Sentiment and Emotions in e-Negotiations

The impact of emotions and their strategic use have been studied from several perspectives. In competitive-cooperative balanced negotiation settings, the strategic expression of anger was found to increase the negotiation partner's concessions and to improve the agreement for the negotiator [3,35]. More generally, emotional expressions can modify a negotiator's perception of the counterpart's priorities and shape the negotiator's strategy [30].

Research into emotions in e-negotiations relies on the linguistic aspects of the expression of emotions. Positive and negative emotions are encoded by the use of positively and negatively associated language cues and their respective frequencies, which represent a certain "tone" or "sentiment" of a message [16]. Business intelligence research is well equipped to study how agents process information conveyed by the language used in textual messages [12]. Textual content provides relevant insights into the position of the message originator through the tone of the language. The subjective tone of text documents can be assessed using so-called sentiment analysis, which relies on computational linguistics. Sentiment analysis measures the positivity or negativity of a text, which in turn reflects affective states of the author or the author's intended display of emotions [25]. Thereby, a higher sentiment score indicates more positive language. Advanced concepts of sentiment analysis allow study of more complex concepts expressed through language, e.g. specific emotions [36].

In the context of e-negotiation, Hine et al. [16] find evidence that e-negotiation participants who reached an agreement show significantly more positive emotion, as measured by sentiment analysis, than those who did not.

2.2 Cultural Differences in Handling Emotions in Negotiations

Culture is commonly defined as patterns of acquired behavior, which are rooted in traditional ideas and related values [17,23,40]. In this section, we outline topics related to the state-of-the-art research being conducted on the role of culture in computer-mediated communication. Research shows that these cultural values and norms can influence negotiation processes and outcomes [22]. Culture affects the display and interpretation of emotions. Kopelman and Rosette [22] contend that cultural background influences the handling of expressed positive or negative emotion in negotiations. For instance, East Asian negotiators, who have a tremendous appreciation for respect and modesty, are more likely to accept offers from negotiation partners who express positive emotion than are Israeli negotiators, who do not value these qualities as highly. The researchers conclude that negotiators' cultural and normative background supplies them with measures by which they assess the display of their counterparts' emotions. Misinterpretation of these emotional expressions can affect the negotiation and damage the relationship. A wide variety of different measures, norms, and values are encountered in cross-cultural negotiations. This increases both the likelihood of misunderstanding and the time needed to negotiate [4]. Adair and Brett [2] observe that intercultural negotiations require more clarifying statements than intracultural negotiations and hence cause more frustration. They suggest that joint gains are higher in intracultural negotiations than in intercultural negotiations because outcomes depend largely on the similarity of the negotiators' understanding of the negotiation, which can differ in international negotiations [37].

2.3 Development of Hypotheses

This study focuses on dyadic e-negotiations. We analyze the interaction of sentiment, culture, and negotiation processes and outcomes.

Sentiment of Previous Message. Reciprocity and affect display are common phenomena in dyadic negotiations [5,8,9]. Although research on reciprocity focuses on the reciprocation of strategies [29], it is likely to assume that there are also effects of reciprocation of emotions, not least because the display of emotions may itself be an effective negotiation strategy [3]. Misunderstanding of a counterpart's emotions may be reciprocated and lead to conflict spirals, which are contentious communication patterns [9]. The occurrence of a conflict spiral makes it difficult for the negotiation to be settled [33] and is linked to hard bargaining and distributive negotiations [9]. George et al. [14] suggest that this behavior in the heat of the moment may also lead to positive spirals and that the effects of reciprocity are different in intercultural negotiations than in intracultural negotiations [1]. Cross-cultural differences in affective reactions and their perception can be reduced to the cultural differences in norms and values, the expression of emotions, and linguistic styles [14]. They argue that affect display is more likely to occur in intercultural negotiations due to the complexity and uncertainty that can accompany cultural differences. Studies show a general tendency towards reciprocity even among trained negotiators [9,44] and reveal differences in reciprocation between intra- and intercultural negotiations [1]. Merging these findings on reciprocity and emotion spirals in inter- and intracultural negotiations, we hypothesize:

Hypothesis 1. *The higher the sentiment of a message, the higher the sentiment of the next message.*

Hypothesis 2. *The impact of a message's sentiment on the next message's sentiment is larger in intercultural than in intracultural negotiations.*

Initiation of the Negotiation. Researchers have showed the importance of the early stages of a negotiation [10]. Many negotiations fail because none of the negotiators takes the initiative to engage their counterpart [13]. This reluctance to initiate a negotiation, which sometimes even entails an element of anxiety [45], is part of many commonly known dysfunctional group phenomena [18]. It has been argued that the likelihood of an individual initiating a negotiation depends on their attitude towards initiation in general [41], which in turn is associated with the individual's beliefs about the appropriateness of doing so, as well as their personality [18]. Both beliefs about appropriateness and personality factors have their roots in the cultural background of the negotiator [10]. Previous research has found direct connections between certain personality factors and initiation [18,41]. Kapoutsis et al. [18] propose the inclusion of culture in the analysis of initiation, since different cultural regions often have different behavioral patterns that could affect the decision to initiate a negotiation [41]. Liu et al. [24] observe that customers from individualistic cultures are more likely to take the initiative to complain when receiving poor service or quality. Volkema and Fleck [41] trace this effect back to the individuals' culturally imprinted perception of appropriate behavior and transfer the findings to the initiation of a negotiation. Researchers also find evidence that women are less likely to initiate negotiations in cultures where women and men are viewed as having substantially different roles in comparison to cultures in which roles are equal [6,41]. Volkema and Fleck [41] suggest that combinations of personality and culture can have a very significant impact on initiation. Their study shows that individuals with a propensity for taking risks, who come from a culture where initiation is seen as appropriate behavior, are more likely to initiate than individuals from cultures which do not support initiation.

Magee et al. [27] report that making the first offer is linked to bargaining advantages. Negotiators who start a negotiation are able to assert their strategy for the entire negotiation, in order to enforce an either integrative or distributive dynamic [32,43]. Thereby, the proactive behavior of the initiator can be interpreted as substantial individual power by the negotiation counterpart [26].

Both strong bargaining power and the appearance thereof result in advantages for the initiator [27]. Hence, we hypothesize:

Hypothesis 3. *The negotiator who initiates the negotiation achieves a higher payoff than their negotiation partner.*

Hypothesis 4. *The cultural background of a negotiator influences the likelihood of being the initiator of the negotiation.*

Hypothesis 5. *The personal conflict-handling style of a negotiator influences their likelihood to be the initiator of the negotiation.*

3   Methods and Data

3.1   Inspire and Descriptive Statistics

For our analyses, we use a dataset of multi-issue, bilateral e-negotiation experiments collected with the web-based negotiation support system Inspire [19]. With Inspire, subjects were randomly designated as one of two parties (a buyer and seller). Participation was anonymous and either party was able to discontinue the negotiation at any time. Negotiations were limited to a fixed duration and the experiments were conducted exclusively in the English language. Each negotiator was able to specify their own preferences and choose strategies, limits, and negotiation styles. They could exchange both offers and messages [16]. Inspire saves the negotiation transcript containing messages, offers, timestamps, and utility scores for both negotiators [20]. In total, our data consists of 1,304 negotiations conducted between 2009 and 2016 and contains 9,924 messages (including offers with messages).

Subjects filled in a pre-negotiation questionnaire providing information regarding their basic demographics along with their personal expectations and objectives concerning the negotiation. The majority of participants indicated an age of 21–25 years (68.25%) and half of the subjects had a background in business and management (50.31%).

In addition, subjects revealed information concerning their personal conflict-handling style by completing the Thomas-Kilmann Conflict Mode instrument (TKI) questionnaire. Kilmann and Thomas [21] developed this instrument to enable the calculation of numerical scores for the five dimensions of interpersonal conflict-handling traits, including *accommodating, compromising, avoiding, collaborating* and *competing*. These conflict-handling traits have been applied in negotiation training to help negotiators develop an awareness of potential bargaining styles [34]. Further studies have showed that TKI is a valid instrument across cultures [15]. Our Inspire dataset offers us TKI scores for all negotiations after 2009, i.e. for 2,225 negotiators. A summary of the most important variables is shown in Table 1. Overall, 81 percent of the negotiations result in an agreement. The average negotiator is an undergraduate in their twenties and a good English speaker.

Table 1. Descriptive statistics for TKI scores and negotiation variables

Negotiation variables	Mean	Standard deviation	Simple correlations				
			(1)	(2)	(3)	(4)	
(1) Agreement	0.81	0.39	1.00				
(2) Level of Age	2.07	0.48	−0.03	1.00			
(3) Level of Education	2.12	0.51	−0.03	0.38	1.00		
(4) Level of English	4.96	1.08	−0.03	0.20	0.32	1.00	
TKIScores	Mean	Standard deviation	Simple correlations				
			(1)	(2)	(3)	(4)	(5)
(1) Accommodating	5.01	2.29	1.00				
(2) Compromising	8.47	2.09	−0.22	1.00			
(3) Avoiding	6.13	2.17	0.00	−0.14	1.00		
(4) Collaborating	5.22	1.87	−0.22	−0.16	−0.37	1.00	
(5) Competing	5.17	2.90	−0.50	−0.34	−0.41	−0.08	1.00

Note: The Agreement category reflects whether the negotiators were able to settle or not (1 for agreement, 0 for no agreement). Level of Age, Education and English are mean values of the negotiators' values. The negotiators are separated into age groups 1–6 (from 20 or less to 51 or more), education groups 1–3 (college, undergraduate, graduate), and English proficiency groups 1–7 (poor to excellent).

3.2 Cultural Variables

To make statements about cultural influences in negotiations, we treat culture as a composite measure of all cultural dimensions within an area and make direct use of the culture-related data given by the demographic categories *country of birth* and *first language* in our data-set. First, we take a look at cultural identity. Rather than a mere country-of-birth approach, we utilize the United Nations geoscheme[1] to sort the countries into larger groups. We apply this geoscheme to the *country of birth* and obtain the sub-regions of East Europe (1,033 observations), West Europe (627), East Asia (366), North America (119) and Other (338). Second, we take language identity into account and group participants by their mother tongue. We restrict this approach to participants who entered only one valid language as their first language. Thereby, we generate the groups of Polish (802), German (429), English (297), Chinese (286) and Other (387). We introduce a dummy variable representing whether the two negotiators' cultural characteristics are different or not (1 for intercultural and 0 for intracultural).

---

[1] World subdivision into macro geographical regions and geographical sub-regions conducted by the UNSD, cf. http://unstats.un.org/unsd/methods/m49/m49regin.htm (12-6-2016).

3.3   Language Sentiment Analysis

Language sentiment analysis measures the positivity or negativity of a text [25]. Positivity or negativity in textual analysis is usually represented by the frequency of words in the text, which are contained within word lists created especially for this purpose (also "dictionaries") [25]. In line with established sentiment analysis methods, we first preprocess our data, i.e. we remove punctuation, numbers and stop words and then stem the remaining words [25]. Then, for each negotiation, we measure the tone or sentiment S_i^j for message i of negotiator $j \in \{1, 2\}$ by scaling the number of positive words p_i^j minus the number of negative words n_i^j by the total word count w_i^j

$$S_i^j = \frac{p_i^j - n_i^j}{w_i^j} \in [-1, 1].$$

We only take into account negotiations in which both negotiators wrote at least one message. This allows us to generate reliable statements demonstrating the effects of sentiment, which leaves us 1,147 negotiations with 9,703 messages.

To calculate the sentiment scores, we make use of the commonly utilized Harvard-IV psychological dictionary of positive and negative words (1,914 positive words and 4,186 negative words) [25,36].

Language, overall, is a very complex and dynamic process of expressing and verbalizing thoughts. Information may not only be transmitted explicitly, but also implicitly in the form of sarcasm or irony, which often differ from country to country or even among regions within a country. Our approach to language sentiment does not account for such covert concepts of embedding information into language. Rather, we rely on the aforementioned concept of language sentiment analysis, which is both established and well respected, in order to approach the role of language sentiment in negotiations with a proven methodology. Furthermore, we do not account for informal concepts of language, such as emoticons or punctuation, which may express implicit messages, e.g. through the use of exclamation marks. While sentiment analysis with a focus on social media, such as Twitter, takes into consideration emoticons and abbreviations [28], we investigate language sentiment in an area closer to business language and thus better represented by a more conservative and more business-language-oriented approach, as represented by the Harvard-IV psychological dictionary [25,36].

4   Results

4.1   Sentiment of Previous Message

We regress the sentiment of a message on the sentiment of the previous message to investigate Hypothesis 1. To overcome heteroscedasticity in the data, we make use of the Newey-West estimator to correct our findings and detect a very significant ($p < 0.001$) positive effect of a message's sentiment on the sentiment of the next message (see Table 2). This effect remains stable when adding further control variables. We do not observe a problem with multicollinearity. In addition, the time between messages, the answer dummy (which represents whether the author of the messages changed), the agreement dummy, the level of age, the level of education, and the level of English do not have a statistically significant influence on the sentiment of a message. To quantify the impact of message sentiment, we standardize our regression coefficient [11] and find that a one standard deviation increase in the sentiment of a message leads to a 20.01% increase in the sentiment of the next message.

With regard to a possible influence of cultural differences on the effect of the sentiment of a message with regard to the sentiment of the next message, we add intercultural dummy variables and their interaction variables with sentiment (i.e. the product of the cultural dummy variables and the sentiment of the previous message). Table 2 reveals a significant positive effect of the intercultural dummy ($\beta_2 = 0.044$, $p < 0.001$) for the country of birth and a significant negative effect of the interaction variable ($\beta_3 = -0.136$, $p < 0.01$), while the effect of the sentiment of the previous message remains significant ($\beta_1 = 0.315$, $p < 0.001$). The impact of message sentiment on the sentiment of the next message is represented by the added

Table 2. Result of OLS regression with Newey-West correction of sentiment of a single message on the sentiment of the previous message with different country of birth's and different first language's dummy and interaction variables

Dependent variable: Sentiment of message				
	(1)	(2)	(3)	(4)
Intercept	0.194*** (42.281)	0.210***(12.722)	0.176***(8.616)	0.174*** (7.907)
Sentiment of previous message	0.208*** (12.727)	0.205*** (12.429)	0.315***(7.953)	0.285*** (6.956)
Days between messages		0.001 (0.883)	0.001(0.802)	0.001 (0.810)
Answer		0.004 (0.598)	0.004 (0.521)	0.005 (0.694)
Agreement		0.011 (1.885)	0.010 (1.719)	0.012 (1.698)
Level of age		−0.006 (−1.157)	−0.004 (−0.750)	−0.003 (−0.565)
Level of education		−0.008 (−1.528)	−0.008(−1.606)	−0.010(−1.826)
Level of English		0.000 (−0.166)	−0.001(−0.553)	0.001(0.242)
Different country of birth			0.044***(3.436)	
Different country of birth × sentiment of previous message			−0.136** (−3.156)	
Different first language				0.036** (2.680)
Different first language × sentiment of previous message				−0.112* (−2.483)
Observations	7,771	7,771	7,754	6,220
R^2	0.041	0.043	0.046	0.044
Adj. R^2	0.041	0.042	0.045	0.042

Stated: Coeff., *t*-Stat. in Parentheses *Significance level:* ***0.001, **0.01, *0.05

coefficient $\beta_1 + \beta_3 = 0.179$ in an intercultural negotiation and by the coefficient $\beta_1 = 0.315$ in an intracultural negotiation, i.e. the impact of the sentiment of a message on the sentiment of the next message is greater in negotiations involving negotiators who were born in the same country than in negotiations in which the negotiators were born in different countries. We receive a similar result regarding first languages, in which the impact of the sentiment of a message on the sentiment of the next message is represented by $\beta_1 + \beta_3 = 0.173$ in intercultural and by $\beta_1 = 0.285$ in intracultural negotiations. Ultimately, both the cultural identity and the linguistic identity approach show that the impact of the sentiment of a message on the sentiment of the next message is greater in intracultural negotiations than in intercultural negotiations. Based on these findings we reject Hypothesis 2.

As is common in language sentiment research, the adjusted R^2 values for the regressions reported in Table 2 are rather small and explain less than five percent of the variance. This comes as no surprise, since textual information, expressed by language, is a very dynamic concept with myriad different possibilities to convey a certain concept (e.g. there are many ways of expressing explicitly or implicitly a disagreement with an offer). The explained variance is similar to other language sentiment research in a business administration and finance context [38,39].

4.2   Initiation of the Negotiation

A first examination of our data shows that the payoff is higher for negotiators who initiate a conversation (see Table 3). Furthermore, we can see that these differences are greater in the lower payoff quantiles than for the negotiators who reach high payoffs.

Table 3. Result of quantile regression of payoff on initiation dummy

Dependent variable: Payoff			
	$Q_{0.20}$	$Q_{0.50}$	$Q_{0.80}$
Intercept	24*** (3.579)	74*** (107.164)	85*** (211.273)
Initiator	23* (2.196)	3*** (3.299)	2** (2.983)

Stated: Coeff., t-Stat. in Parentheses
Significance level: ***0.001, **0.01, *0.05
Observations: 2,520

We regress the payoff scores on the initiation dummy and correct our results with the Newey-West estimator to overcome autocorrelation in our data. Thereby, we confirm the results by adding the TKI measures (*accommodating, compromising, avoiding* and *collaborating*) to the regression as independent variables. We leave out *competing* because of collinearity with the other TKI scores. Linearly regressing payoff on the initiation dummy and the four TKI scores shows a positive effect of initiation on payoff (see Table 4), thus supporting Hypothesis 3. Incidentally, we do not observe a significant effect of the TKI dimensions on payoff.

Table 4. Result of Newey-West corrected OLS regression of negotiation payoffs

Dependent variable: Payoff of Negotiator					
	(1)	(2)	(3)	(4)	(5)
Intercept	62.009*** (54.194)	65.686*** (34.974)	62.474*** (15.134)	59.937*** (12.406)	56.546*** (7.913)
Initiator	2.979*** (3.852)	2.855*** (3.797)	2.751*** (3.605)	2.758*** (3.619)	2.768*** (3.648)
Accommodating		−0.722* (−2.268)	−0.655* (−2.025)	−0.646* (−2.006)	−0.574 (−1.68)
Compromising			0.346 (0.951)	0.398 (1.076)	0.475 (1.197)
Avoiding				0.334 (1.046)	0.447 (1.205)
Collaborating					0.322 (0.745)
Observations	2,122	2,122	2,122	2,122	2,122
R^2	0.002	0.005	0.005	0.006	0.006
Adj. R^2	0.002	0.004	0.004	0.004	0.004

Stated: Coeff., t-Stat. in Parentheses
Significance level: ***0.001, **0.01, *0.05

Table 5. Results of logit regressions of the initiation dummy

Dependent variable: Initiation			
Area of birth		First language	
Intercept	−0.406 (−0.905)	Intercept	−0.231 (−0.494)
East Asia	0.216 (1.218)	Chinese	−0.041 (−0.222)
East Europe	0.316* (2.189)	English	−0.090 (−0.485)
North America	−0.123 (−0.395)	German	0.395* (2.399)
West Europe	0.487** (3.107)	Polish	0.188 (1.315)
Age Group	−0.090 (−1.282)	Age group	−0.125 (−1.677)
Accommodating	−0.012 (−0.548)	Accommodating	−0.001 (−0.047)
Compromising	0.060** (2.573)	Compromising	0.053* (2.131)
Avoiding	−0.018 (−0.760)	Avoiding	−0.016 (−0.643)
Collaborating	−0.012 (−0.443)	Collaborating	0.003 (0.104)
Observations	2,005	Observations	1,757

Stated: Coeff., z-Stat. in Parentheses
Significance level: ***0.001, **0.01, *0.05

To test Hypotheses 4 and 5, we logistically regress the initiation dummy on birth country group dummies and TKI scores. The results in Table 5 reveal a significant positive effect of East Europe group ($p < 0.05$), West Europe group ($p < 0.01$) and *compromising* ($p < 0.01$), i.e. negotiators from one of these regions and highly compromising negotiators were more likely to initiate their negotiation than negotiators from other regions. West Europe turns out to be the strongest predictor in favor of initiation among the areas of birth. We logistically regress on dummies of first languages of the negotiators and TKI scores. A significant positive effect of German group ($p < 0.05$) and a significant positive coefficient of *compromising* ($p < 0.05$) are observed (see Table 5). Hence, the negotiators whose first language is German, as well as those highly willing to compromise, were more likely to initiate than other negotiators. All in all, we found evidence that both negotiators' cultural and linguistic identities, along with their personal conflict-handling style. influence the likelihood of their initiating a negotiation. Thus, Hypotheses 4 and 5 are supported.

A potential limitation of this approach is that the negotiation starting time is always midnight GMT, which might interfere with the geographically defined cultural groups. We replicated our analysis and excluded all negotiations starting less than nine hours after the negotiation opening (a total of 30 negotiations for the area-of-birth approach and 27 for the first-language approach were excluded). Our results remain robust against such a variation.

5 Conclusion

From our analyses, we are able to determine that reciprocation of sentiment is common in both inter- and intracultural e-negotiations. Contrary to our presumption, this effect is larger in intracultural negotiations than in intercultural negotiations. This could be due to a more cautious communication style, including more clarifying statements, which negotiators presumably apply if they encounter differences in perceptions, emotions, and communication patterns [2].

We show that initiation increases payoffs. This effect is especially pronounced with respect to low payoffs. We conclude that by initiating the negotiation, negotiators are able to minimize the risk of low payoffs. Subjects from certain cultural regions, like German-speaking Western Europe, emerge as especially likely to initiate a negotiation, which is consistent with previous studies showing that culture and personality influence initiation behavior [18,41]. However, German-speaking Western Europeans also constitute one of the largest groups of participants in our dataset. Hence, it would be interesting to confirm these findings with other datasets and regions.

Further research could also classify cultures differently than merely by country, region, or language. For example, Hofstede's culture dimensions could be used to explain why some cultures are more likely to initiate than others. This is an extension of our work into which we plan to conduct further research, as this paper focuses more on personality traits according to TKI.

References

1. Adair, W.L.: Exploring the norm of reciprocity in the global market: U.S. and Japanese intra- and inter-cultural negotiations. Acad. Manag. Proc. 1999(1), A1–A6 (1999)
2. Adair, W.L., Okumura, T., Brett, J.M.: Negotiation behavior when cultures collide: the United States and Japan. J. Appl. Psychol. 86(3), 371–385 (2001)
3. Adam, H., Brett, J.M.: Context matters: the social effects of anger in cooperative, balanced, and competitive negotiation situations. J. Exp. Soc. Psychol. 61, 44–58 (2015)
4. Adler, N.J., Gundersen, A.: International Dimensions of Organizational Behavior. South-Western Cengage Learning, Mason (2008)

5. Axelrod, R.: The Evolution of Cooperation. Basic Books, New York (1984)
6. Babcock, L., Gelfand, M.J., Small, D., Stayn, H.: Gender differences in the propensity to initiate negotiations. In: de Cremer, D., Zeelenberg, M., Murnighan, J.K. (eds.) Social Psychology and Economics, pp. 239–259. Lawrence Erlbaum, Mahwah (2006)
7. Baron, R.A., Fortin, S.P., Frei, R.L., Hauver, L.A., Shack, M.L.: Reducing organizational conflict: the role of socially-induced positive affect. Int. J. Confl. Manag. **1**(2), 133–152 (1990)
8. Boulding, K.: Conflict and Defense: A General Theory. Harper & Brothers, New York (1962)
9. Brett, J.M., Shapiro, D.L., Lytle, A.L.: Breaking the bonds of reciprocity in negotiations. Acad. Manag. J. **41**(4), 410–424 (1998)
10. Brett, J.M.: Negotiating Globally: How to Negotiate Deals, Resolve Disputes, and Make Decisions Across Cultural Boundaries. The Jossey-Bass Business & Management Series, 2nd edn. Jossey-Bass, San Francisco (2007)
11. Bring, J.: How to standardize regression coefficients. Am. Stat. **48**(3), 209 (1994)
12. Chen, H., Chiang, R.H.L., Storey, V.C.: Business intelligence and analytics: from big data to big impact. MIS Q. **36**(4), 1165–1188 (2012). http://dl.acm.org/citation.cfm?id=2481674.2481683
13. Curhan, J.R., Pentland, A.: Thin slices of negotiation: predicting outcomes from conversational dynamics within the first 5 minutes. J. Appl. Psychol. **92**(3), 802–811 (2007)
14. George, J.M., Jones, G.R., Gonzalez, J.A.: The role of affect in cross-cultural negotiations: an integrative overview. J. Int. Bus. Stud. **29**(4), 749–772 (1998)
15. Herk, N.A., Thompson, R.C., Thomas, K.W., Kilmann, R.H.: International technical brief for the Thomas-Kilmann conflict mode instrument (2011). https://www.cpp.com/contents/tki_research.aspx
16. Hine, M.J., Murphy, S.A., Weber, M., Kersten, G.: The role of emotion and language in dyadic E-negotiations. Group Decis. Negot. **18**(3), 193–211 (2009)
17. Hofstede, G.: Culture's Consequences: Comparing Values, Behaviors, Institutions, and Organizations Across Nations. Sage, London (2001)
18. Kapoutsis, I., Volkema, R.J., Nikolopoulos, A.G.: Initiating negotiations: the role of machiavellianism, risk propensity, and bargaining power. Group Decis. Negot. **22**(6), 1081–1101 (2013)
19. Kersten, G.E., Noronha, S.J.: WWW-based negotiation support: design, implementation, and use. Decis. Support Syst. **25**(2), 135–154 (1999)
20. Kersten, G.E., Zhang, G.: Mining inspire data for the determinants of successful internet negotiations. Cent. Eur. J. Oper. Res. **11**(3), 297–316 (2003)
21. Kilmann, R.H., Thomas, K.W.: Developing a forced-choice measure of conflict-handling behavior: the mode instrument. Educ. Psychol. Meas. **37**(2), 309–325 (1977)
22. Kopelman, S., Rosette, A.S.: Cultural variation in response to strategic emotions in negotiations. Group Decis. Negot. **17**(1), 65–77 (2008)
23. Kroeber, A.L., Kluckhohn, F.: Culture: a critical review of concepts and definitions. In: Papers of the Peabody Museum of Archaeology and Ethnology, vol. 47, no. 1. Harvard University (1952)
24. Liu, B.S., Furrer, O., Sudharshan, D.: The relationships between culture and behavioral intentions toward services. J. Serv. Res. **4**(2), 118–129 (2001)
25. Loughran, T., McDonald, B.: Textual analysis in accounting and finance: a survey. J. Account. Res. **54**(4), 1187–1230 (2016)

26. Magee, J.C.: Seeing power in action: the roles of deliberation, implementation, and action in inferences of power. J. Exp. Soc. Psychol. **45**(1), 1–14 (2009)
27. Magee, J.C., Galinsky, A.D., Gruenfeld, D.H.: Power, propensity to negotiate, and moving first in competitive interactions. Personal. Soci. Psychol. Bull. **33**(2), 200–212 (2007)
28. Pak, A., Paroubek, P.: Twitter as a corpus for sentiment analysis and opinion mining. In: Proceedings of the Seventh Conference on International Language Resources and Evaluation (2010)
29. Parks, C.D., Komorita, S.S.: Reciprocity research and its implications for the negotiation process. Int. Negot. **3**, 151–169 (1998)
30. Pietroni, D., van Kleef, G.A., de Dreu, C.K., Pagliaro, S.: Emotions as strategic information: effects of other's emotional expressions on fixed-pie perception, demands, and integrative behavior in negotiation. J. Exp. Soc. Psychol. **44**(6), 1444–1454 (2008)
31. Pruitt, D.G., Carnevale, P.J.: Negotiation in Social Conflict. Open University Press, Buckingham (1993)
32. Putnam, L.L.: Reframing integrative and distributive bargaining: a process perspective. In: Sheppard, B.H., Bazerman, M.H., Lewicki, R.J. (eds.) Research on Negotiation in Organizations, vol. 2, pp. 3–30. JAI Press, Greenwich (1990)
33. Schelling, T.C.: The Strategy of Conflict. Harvard University Press, Cambridge (1960)
34. Shell, G.R.: Teaching ideas: bargaining styles and negotiation: the Thomas-Kilmann conflict mode instrument in negotiation training. Negot. J. **17**(2), 155–174 (2001)
35. Sinaceur, M., Tiedens, L.Z.: Get mad and get more than even: when and why anger expression is effective in negotiations. J. Exp. Soc. Psychol. **42**(3), 314–322 (2006)
36. Stone, P.J.: General inquirer Harvard-IV dictionary (2002). http://www.wjh.harvard.edu/~inquirer/. Accessed 9 Dec 2002
37. Swaab, R., Postmes, T., Neijens, P.: Negotiation support systems: communication and information as antecedents of negotiation settlement. Int. Negot. **9**(1), 59–78 (2004)
38. Tetlock, P.C.: Giving content to investor sentiment: the role of media in the stock market. J. Financ. **62**, 1139–1168 (2007)
39. Tetlock, P.C., Saar-Tsechansky, M., Macskassy, S.: More than words: quantifying language to measure firms' fundamentals. J. Financ. **63**, 1437–1467 (2008)
40. Tylor, E.B.: Primitive Culture: Researches into the Development of Mythology, Philosophy, Religion, Art, and Custom. Gordon Press, New York (1974)
41. Volkema, R.J., Fleck, D.: Understanding propensity to initiate negotiations. Int. J. Confl. Manag. **23**(3), 266–289 (2012)
42. Weingart, L., Smith, P., Olekalns, M.: Quantitative coding of negotiation behavior. Int. Negot. **9**(3), 441–456 (2004)
43. Weingart, L.R., Olekalns, M.: Communication processes in negotiation: frequencies, sequences, and phases. In: Gelfand, M.J., Brett, J.M. (eds.) The Handbook of Negotiation and Culture, pp. 143–157. Stanford Business Books, Stanford (2004)
44. Weingart, L.R., Prietula, M.J., Hyder, E.B.: Knowledge and the sequential processes of negotiation: a Markov chain analysis of response-in-kind. J. Exp. Soc. Psychol. **35**, 266–393 (1999)
45. Wheeler, M.: Anxious moments: openings in negotiation. Negot. J. **20**(2), 153–169 (2004)

Nucleolus-Based Compensation Payments for Automated Negotiations of Complex Contracts

Gabriel Guckenbiehl$^{(\boxtimes)}$ and Tobias Buer

Computational Logistics Junior Research Group,
University of Bremen, Bibliothekstr. 1, 28359 Bremen, Germany
{guckenbiehl,tobias.buer}@uni-bremen.de

Abstract. Automated negotiation mechanisms (ANMs) can be used to semi-automatically negotiate well-structured but complex contracts. Such negotiations among multiple negotiators, who are represented by software agents, can easily reach a deadlock because they block each others proposals. This of course leads to inferior results. Our aim is to improve the performance of ANMs. In this paper, we try to overcome deadlocks during automated single negotiation text by using compensation payments. Compensation payments are calculated by a mediator according to the Nucleolus method from cooperative game theory. It guarantees that a unique payment is calculated in any case. Furthermore, it lies in the Core, if the Core exists. The proposed ANM can be vulnerable against shading the desired compensation payments. However, our computational experiments suggest that the negotiation results are both superior and faster compared to an ANM without compensation payments. The dominance increases with an increasing number of negotiators.

Keywords: Automated negotiation · Negotiation deadlock · Compensation payment · Cooperative game theory · Nucleolus

1 Introduction

Automated negotiation mechanisms (ANMs) can support businesses to coordinate their inter-organizational planning decisions. Inter-organizational business systems, e.g. supply chains, have to be coordinated in order to improve effectiveness and efficiency of the entire system. Usually, coordination takes place at the planning level. Real-world planning problems are often extensive and include many interdependent decisions [2,4]. They are often modeled as NP-hard problems which are computationally difficult, even for a single decision-maker who does not even require to coordinate decisions. On the one hand, the group of involved business partners is interested in a competitive overall system, i.e., they want to maximize social welfare. On the other hand, the business partners (referred to as agents) represent independent companies who are self-interested, autonomous and have private information.

M. Schoop and D.M. Kilgour (Eds.): GDN 2017, LNBIP 293, pp. 145–157, 2017.
DOI: 10.1007/978-3-319-63546-0_11

We use a stylized optimization problem in order to study ANMs. In this way we hope to avoid influences on the ANM performance which are introduced by using problem specific heuristics. Those heuristics exploit specific structures of the problem at hand in an intelligent way. In a singular case this seems to be the way to go. However, in order to create results which are more independent from specific NP-hard optimization problems and related heuristics we use a plain but difficult negotiation problem. As in [12], we assume a mediated single negotiation text of complex contracts with multiple, pairwise interdependent issues. A contract c includes n clauses. If clause $i = 1, \ldots, n$ is active in contract c, then $c_i = 1$, otherwise $c_i = 0$. Let A be a set of agents who negotiate. Each agent $a \in A$ has an influence matrix H^a to measure the agent's utility for the pairwise presence of the n clauses. Agent a's total utility of contract c is calculated according to (1).

$$U^a(c) = \sum_{i,j} H^a_{ij} \cdot c_i \cdot c_j \tag{1}$$

If the clauses i and j are both active in contract c, the utility of agent $a \in A$ increases by H^a_{ij}. In the stylized problem, only two clauses are interdependent, respectively. Exploring three or more interdependent clauses would present significant challenges on the computational effort as well as the available input data. For all agents, the values of H^a_{ij} are randomly chosen from $[-100, +100]$.

These characteristics easily lead to local optimal contracts ("deadlocks") during the negotiation process. To overcome deadlocks during automated negotiations, Klein et al. [12] propose a stochastic acceptance criterion inspired by the simulated annealing metaheuristic. Agents may accept a contract proposal even if it is inferior. This concept has been successfully implemented in distributed planning problems related to production planning in supply chains [3,6,16]. Another approach to overcome deadlocks is negotiating multiple contracts simultaneously [8]. As a result of both approaches, deadlocks are overcome more easily and the ANMs find contracts with a significantly higher social welfare. However, another way to overcome deadlocks in ANMs may be to use compensation payments. In inter-organizational business systems the agents are heterogenous with respect to their value contribution and economic relevance for the system. Think of a world-wide automobile supply chain, for example. There is a large original equipment manufacturer and major suppliers, but also many small and medium sized suppliers. Given such a diverse group of agents, a contract proposal can easily increase social welfare but is still rejected by some agents because their utility declines. However, the increase in social welfare might be that strong so that the agents whose utility increase are able to compensate the remaining agents for their declines. We assume transferable utility and call these payments compensation payments.

Our aim is to create an ANM that arrives at global superior contracts. We try to overcome deadlocks during automated negotiations by using compensation payments. Therefore, we present an ANM for mediated single negotiation text with integrated compensation payments based on the Nucleolus solution concept from cooperative game theory. We show that the additional information and incentives used in this new mechanism lead to superior solutions in one-tenth of the negotiation rounds. Furthermore, the results are much more robust when the number of negotiators increases.

The remaining paper is organized as follows. Section 2 gives a short overview of the literature. The proposed ANM with Nucleolus-based compensation payments is introduced in Sect. 3. Its performance is evaluated by means of computational experiments in Sect. 4. Section 5 concludes the paper.

2 Review of the Literature

Generic Mechanisms. Negotiations are necessary due to the agents' unwillingness to share their private information which makes it impossible to employ a central planning approach. Lopes et al. [14] provide an overview about research in the field of automated negotiation. Jennings et al. [9] present a generic framework which helps to differentiate and analyze negotiation techniques. A generic agent architecture for multi attribute negotiations is presented by Jonker et al. [10]. Klein et al. [13] analyze the state of the art in negotiation theory and give insights on how negotiation methods can be used for collaborative design. Finally, Klein et al. [12] use a stochastic acceptance criterion from the simulated annealing metaheuristic in order to overcome deadlocks. They compare their approach to pure hill climbing agents who get stuck easily and study the resulting prisoner's dilemma. They also present some techniques to mitigate this dilemma.

Compensation Payments. Compensation payments (also often referred to as side payments) are a concept from cooperative game theory and emerge from the question how a jointly generated utility should be distributed among the members of a coalition. Members of a coalition are also referred to as players or, in the context of this paper, as cooperating agents. Occasionally, compensation payments are used during collaborative lot size planning. For example, Shapley-based compensation payments are used in [5]. A drawback of that approach in the context of inter-organizational lot sizing is the need to reveal many private information in the form of marginal values of contract variants.

Compensation payments are also successfully used without justifying them by cooperative game theory approaches. Dudek and Stadtler [4] consider a two agent scenario where the negotiation proceeds as long as social welfare increases. Compensation payments are calculated based on a goal-programming approach. The allocation of payments among the agents is not discussed, because it does not arise in bilateral negotiations. In the mechanism presented in [7] the compensation payments are subject to fine-grained negotiations among multiple agents.

3 Negotiation Protocol

Three variants of automated negotiation approaches are studied. They are referred to as HC, SA, and NP. Just as Klein et al. [12] we study single negotiation text and two types of agents: hill climber agents and annealer agents. A hill climber agent accepts a new contract proposal only if it increases the agent's utility. In contrast, an annealer agent accepts a proposal depending on a stochastic acceptance criterion which depends on (a) the agent's utility of the proposal and (b) on the progress of the negotiation. We denote the ANM using only hill climber agents as hill climbing ANM (HC) and the ANM using only simulated annealing agents as simulated annealing ANM (SA).

For the third variant, we propose an ANM based on compensation payments. We also assume hill climber agents who consider their utility of a proposal and take compensation payments by other agents into account. These compensation payments are computed by the mediator. The compensation payments conform with the Nucleolus properties known from cooperative game theory. Therefore, we denote this approach as Nucleolus-based compensation payments ANM (NP).

3.1   Overview of the Protocol

An overview of the negotiation protocol is shown by Fig. 1. The negotiation consists of two stages. Stage 1 is used by HC, SA, and NP. In contrast, Stage 2 is only used by NP. Contract c is the initial contract accepted by all agents. The mediator starts the negotiation in Stage 1 by announcing a modified contract proposal c' (see Sect. 3.2). This contract is evaluated by the agents and each agent decides whether to accept or reject c' (see Sect. 3.3). If all agents vote to accept c', then $c \leftarrow c'$. For HC and SA, the next round starts with c by repeating Stage 1. The process iterates until r^{max} rounds are over.

For NP and the case that at least one agent vetoes the new proposal c', Stage 2 begins. In Stage 2 compensation payments are calculated (see Sect. 3.4). Therefore, the agents reveal their utility difference ΔU^a (stated in monetary units). That is, (1) How much money does agent a desire in order to accept c'? Or, alternatively: (2) How much money is agent a able to waive so that c' is mutually accepted? Based on this data, compensation payments are calculated by the mediator. If feasible payments are possible, then $c \leftarrow c'$, otherwise c is not updated. In any case, a new negotiation round starts with Stage 1.

3.2   Modification of a Contract Proposals by the Mediator

During the negotiation the mediator is the only entity who modifies a contract proposal. The negotiation always starts with the proposal where all clauses are set to zero. In each round, the mediator modifies the currently accepted contract c and proposes a new contract c'. Therefore, exactly one of the n clauses of the binary contract c is flipped. That is, an integer i between 1 and n is chosen randomly and the value of the binary variable c_i' is inverted. The new proposal c' is passed on to all agents.

3.3   Acceptance of New Contract Proposals

Each agent evaluates the mediator's modified contract proposal c'. The change in agent a's utility is given by $\Delta U^a = U^a(c') - U^a(c)$.

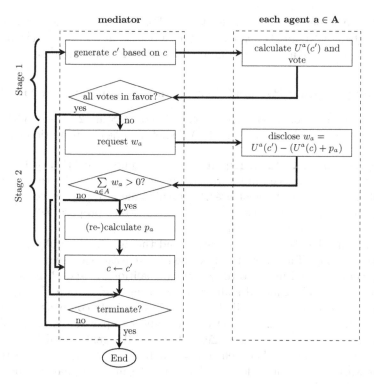

Fig. 1. Overview of the automated negotiation protocol; compensation payments are $p_a = 0$ in round $r = 0$.

- A hill climber agent a in HC accepts a proposal if it increases the agent's utility, i.e., $\Delta U^a \geq 0$.
- A compensation agent a in NP accepts a proposal if it increases the agent's utility considering the compensation payments p^a, i.e., $\Delta U^a + p^a \geq 0$. Payments p^a may be positive or negative and we assume utility is measured in monetary units. For $p^a = 0$ a compensation agent acts like a hill climber agent.
- An annealer agent a in SA uses a stochastic acceptance criterion. The probability that agent a accepts a contract is $\max(1.0, e^{\Delta U^a/T_r})$. For $\Delta U^a \geq 0$ a contract proposal is always accepted. For $\Delta U^a < 0$ it depends on the parameter T_r which is a measure for the progress of the negotiation.
  T_r refers to a so-called temperature parameter which takes up a metaphoric reference to the annealing process of heated metal. This inspired the meta-heuristic simulated annealing which is very successful for solving NP-hard combinatorial optimization problems [11]. Long story short: In each round the temperature decreases. The automated negotiation process is scheduled for r^{max} rounds. The temperature is updated in each round. In round r the temperature T_r is $T_r = \frac{1}{10}(r^{max} - r)$. For this reason, the probability for an agent a to accept an inferior contract decreases the longer the negotiation lasts and the stronger negative ΔU^a is.

The acceptance decision completes Stage 1. For HC and SA a new negotiation round starts. For NP and the case that at least one agent refuses the new proposal c', compensation payments are computed in Stage 2 which are described next.

3.4 Compensation Payments According to Aumann and Maschler

Possible compensation payments are calculated in Stage 2 of the ANM. This stage exists only in NP. Both HC and SA do not include compensation payments.

Willingness to Pay. Stage 2 requires that the agents reveal their willingness to pay $w_a, a \in A$. We proceed in the remainder of the paper under the simplifying assumption that for each agent $a \in A$ the willingness to pay w_a equals exactly the difference of the utility values of the new contract proposal $U^a(c')$ and the already established contract $U^a(c)$.

We assume that there is no cheating and the agents truthfully reveal their willingness to pay. If a contract c' is vetoed by at least one agent, the coalition A of agents is divided into two (disjoint) subcoalitions F and R. F includes all agents in favor of c'. R includes all agents who reject c'. Group F can convince group R to accept contract c' if they are able to jointly compensate the losses of the agents in R. We define agent a's willingness to pay as the marginal value of a contract proposal, i.e. $w_a = \Delta U^a$. If $w_a \geq 0$, agent a still benefits from c', if a pays any amount between 0 and w_a. If $w_a < 0$, then agent a requires a compensation payment at least as large as $|w_a|$ in order that a is better off. Therefore, F is able to compensate R, if

$$\sum_{a \in A} w_a \geq 0. \tag{2}$$

Otherwise, a compensation is not possible and Stage 2 terminates. The follow up question is: If a compensation is possible, then how much should each agent pay or receive? The allocation of joint profits is the subject of solution concepts from cooperative game theory.

Core and Nucleolus. A solution of a cooperative game is an imputation $x = (x_1, \ldots, x_m)$. That is, x describes the amount of money each agent pays or receives. Well-known solution concepts are the Core and the Nucleolus. The Core describes a set of imputations. All imputations within the Core are non-dominated by any other imputation. An imputation within the Core cannot be blocked by any other feasible imputation; there is no subcoalition of agents who can do better when leaving the grand coalition. This is a useful property for automated negotiations of multiple agents because we can focus on contracts that include all agents. It is not necessary to negotiate contracts for subsets of agents because for any subset of agents there is no incentive to leave the grand coalition A. Disadvantages of the Core in the context of ANM are:

- The Core describes a set of imputations. The mediator has to select a single imputation in order to suggest actual compensation payments.

- The Core may be empty. Then no compensation payments are suggested at all. This might be too drastic, because a non-satisfying payment might be better for the negotiation progress than no payment.
- The most serious disadvantage in an ANM context with asymmetric information appears to be the fact that values of all subcoalitions have to be revealed in order to calculate the Core.

Some of the drawbacks are addressed by a concept called the Nucleolus introduced by Schmeidler in [15]. It is achieved by lexicographically minimizing the incentive of each subcoalition to leave the grand coalition. Therefore, by construction, it consists of a single imputation, which is within the core if one exists and a close approximation if not.

For most cooperative games, calculating the Nucleolus is computationally more challenging than computing the Core. It also requires information about the values of subcoalitions which are usually private or even unknown in the studied negotiation context. However, Aumann and Maschler [1] showed how to easily compute a Nucleolus imputation for the situation where a subcoalition F is able to compensate a subcoalition R $(F, R \subset A)$. This fits the situation at hand well.

Calculation of Nucleolus-Based Compensation Payments. Those agents $a \in A$ with a negative willingness to pay $w_a < 0$ have to be compensated in order to accept proposal c'. The question is, how much should each agent with a positive willingness to pay actually pay? Aumann and Maschler [1] proposed an algorithm to compute unique payments in bankruptcy games which fit the problem at hand. Let the components of the willingness to pay vector w of the m agents be sorted in ascending order such that $w_a \leq w_{a+1}$:

$$w_1 \leq w_2 \leq \ldots \leq w_{s-1} \leq w_s \leq w_{s+1} \leq \ldots \leq w_m,$$

where $w_{s-1} \leq 0$ and $w_s \geq 0$. The subcoalition $F \subset A$ of agents in favor contains the agents s to m. They can use $\sum_{a<s}^{m} w_a$ monetary units to compensate the agents 1 to $s-1$ which form subcoalition $R \subset A$ that refuses contract c'. Agents in R receive the following payments:

$$p_a = |w_a| \quad \forall a \in R \subset A. \tag{3}$$

The procedure of Aumann and Maschler [1] is used to calculate the actual payments of the agents in F who are in favor of c'. The Algorithm 1 distinguishes two cases. In the first case, the total compensation $C_R = \sum_{a=1}^{s-1} |w_a|$ demanded by the agents in R is less than half the total willingness to pay $W_F = \sum_{a=s}^{m} w_a$ of the agents in F. The amount to be paid by each agent is increased incrementally. Each agent in $s, \ldots, m$ pays $w_s/2$. Afterwards, agent s is excluded. Then, the remaining money is split among the remaining agents $s+1, \ldots, m$ until each agent pays $w_{s+1}/2$. This process terminates after agent m has been considered. The second case considers $C_R > W_F/2$. That is, in an analogous way, the amount to be kept by each agent is increased incrementally.

input : s, m, willingness to pay w, total compensation $C_R = \sum_{a=1}^{s-1} |w_a|$
output: Payments p_a of the agents $a = s, \ldots, m$ in favor

if $C_R > \sum_{a=s}^{m} \frac{w_a}{2}$ then
$\quad | \quad C \leftarrow \sum_{a=s}^{m} \frac{w_a}{2} - C_R;$
else
$\quad \lfloor \quad C \leftarrow C_R;$

$p_a \leftarrow 0, \forall a = s, \ldots, m;$
for $a \leftarrow s$ to m do
$\quad | \quad$ if $C > (m-a)\frac{w_a}{2}$ then
$\quad\quad | \quad d \leftarrow \frac{w_a}{2} - p_a;$
$\quad\quad | \quad$ for $j \leftarrow a$ to m do
$\quad\quad\quad \lfloor \quad p_j \leftarrow p_j + d \; ;$
$\quad\quad | \quad C := C - (m-a) \cdot d \; ;$
$\quad | \quad$ else
$\quad\quad | \quad$ for $j \leftarrow a$ to m do
$\quad\quad\quad \lfloor \quad p_j \leftarrow p_j + C/(m-j+1) \; ;$

if $C_R > \sum_{a=s}^{m} \frac{w_a}{2}$ then
$\quad | \quad$ for $a \leftarrow s$ to m do
$\quad\quad \lfloor \quad p_a \leftarrow w_a - p_a;$

Algorithm 1. Calculation of compensation payments.

Aumann and Maschler [1] showed that this approach leads to a unique imputation which is identical to the Nucleolus solution of the cooperative game at hand. A drawback of this approach is the required truthfulness of the agents. The agents have to reveal their willingness to pay. However, if an agent in F or R understates its willingness to pay – that is, $w_a < \Delta U^a$ – the chance increases that the proposal might not be jointly accepted. This would be a drawback for the untruthful agent a. Overstating the willingness to pay for an agent in R is irrational. However, an agent in F might overstate its willingness to pay if (a) without overstating the contract would not be jointly accepted and (b) the actual payment p_a computed by Algorithm 1 would still be below ΔU^a. This manipulation requires very detailed information about the other agents and the mechanism. Furthermore, it appears as a slight manipulation only. For the computational benchmark experiments, we assume truthful behavior for both ANMs NP and SA.

Magnitude of Revealed Information. As the agents have to reveal their willingness to pay for some contract proposals the question arises, whether a mediator would be able to re-engineer the utility function of the agents. For this, one must derive the entries of the upper triangle matrix obtained by folding the influence matrix alongside the main diagonal. From algebra one can derive that to determine the entries of an $n \times n$ upper triangle matrix one needs an equality-system of $\frac{n^2+n}{2}$ equalities, i.e., the number of its non-zero entries. To set up one of these equalities a new disclosure of willingness to pay is needed.

Therefore information must be disclosed no less than $\frac{n^2+n}{2}$ times to re-engineer the utility function.

Negotiation continues after the best solution is found. Afterwards, due to the chosen neighborhood function, there will be no more than n distinct contract proposals. Therefore, as long as the best solution is found in less than $\frac{n^2-n}{2}$ rounds the mediator is not able to re-engineer the utility function. As our computational experiments in Sect. 4.3 show, this might become an issue for very small instances with only 25 clauses but not for bigger instances.

4   Discussion of Computational Results

4.1   Experimental Setup

We use computational experiments to evaluate the performance of the NP. We have randomly created six sets of benchmark instances with 2, 5, and 10 agents as well as 25 and 100 contract clauses, respectively. Each set includes 100 random instances. Eventually, an instance is defined by an influence matrix H^a for each agent a, see [12]. H^a is a matrix whose values are drawn randomly from the integer interval $[-100, 100]$. We refer to I^2_{100}, I^5_{100}, and I^{10}_{100} as the sets with two, five and ten agents who negotiate a contract with 100 clauses. Analog, I^2_{25}, I^5_{25}, and I^{10}_{25} denotes the instance sets with 25 clauses and 2, 5, and 10 agents, respectively. Instances with 25 clauses are also referred to as "small instances".

We solve all instances via a central planning approach in order to get strong reference values of the attainable social welfare. For the small instances with 25 clauses we were able to compute the maximum social welfare via the commercial solver "IBM ILOG CPLEX Optimization Studio 12.6". For these instances the social welfare is the actual optimum. This is a true upper bound for all results achievable by any ANM. For the large instances with 100 clauses, we implemented a variant of SA with only one agent as a central planning approach (CSA). CSA approximates the maximum social welfare. This is a strong reference value. However, because it is only an approximation there is no guarantee that it is a true upper bound for the presented ANMs in this study.

4.2   Quality of Automated Negotiation Mechanisms

The quality of the ANMs is measured in terms of their ability to maximize social welfare. Figures 2 and 3 show the aggregated results for the small and the large instance sets. The results are given relative to the social welfare reference value.

Contracts found by HC are clearly outperformed by NP and SA. The more agents participate in an negotiation, the inferior the contracts of HC become. For the 10 agents instances I^{10}_{25} and I^{10}_{100} the median is zero. That is, no contract proposals are found which allow every agent to increase its utility. Frequently, the literature studies ANMs for two agents. Although it might be straight forward to generalize them to the multi-agent case, the performance might significantly deteriorate. While for some applications the results achieved by HC on the two

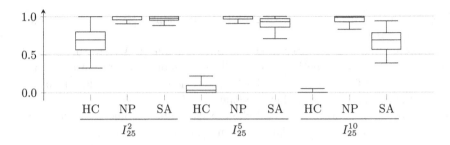

Fig. 2. Boxplots for negotiations of two, five, and ten agents on the set of 100 small instances with 25 clauses each.

instance case might be sufficient, HC is simply not applicable to the five and ten agent cases.

On the other hand, NP and SA perform significantly better for the five and ten instance cases. With respect to maximizing social welfare, NP outperforms SA. As Fig. 3 shows, the medians of the contracts negotiated by NP are very close to the reference social welfare. The advantage of NP over SA grows with more negotiating agents. An explanation could be that the compensation payments are much more targeted with respect to the search direction than the randomized acceptance function of SA. However, the agents have to reveal their willingness to pay in NP which is not required by SA. Comparing NP and SA on the given instance sets it becomes obvious, that NP is also the more robust ANM. The whiskers for NP are considerably closer to their respective medians compared to SA, see Figs. 2 and 3.

A pairwise comparison of all ANMs is shown in Table 1. CSA refers to the central planning approach, i.e. SA with only one agent. The table on the left states the results for the 100 instances of set I_{100}^2. In the middle, the results for the I_{100}^5 instance set are shown and the rightmost table shows the results for I_{100}^{10}. Table 1 indicates how often an ANM in a row is (strictly) outperformed by an ANM in a column. For example, on the 100 instances of set I_{100}^5 the mechanism NP outperforms SA 94 times. Another example: In the 100 instances of set

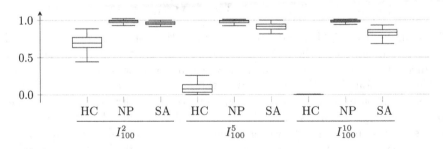

Fig. 3. Boxplots for negotiations of two, five, and ten agents on the set of 100 large instances with 100 clauses each.

Table 1. Number of cases (out of 100) in which one negotiation mechanism (column) provides actually better results than another (row).

I_{100}^2	HC	NP	SA
CSA	0	17	1
HC	-	100	100
NP	0	-	24
SA	0	76	-

I_{100}^5	HC	NP	SA
CSA	0	9	1
HC	-	100	100
NP	0	-	6
SA	0	94	-

I_{100}^{10}	HC	NP	SA
CSA	0	7	0
HC	-	100	100
NP	0	-	0
SA	0	100	-

I_{100}^{10} the mechanism NP outperforms the central planning CSA 7 times. This is possible because we use heuristic reference values for the large instances.

Altogether, the integration of compensation payments based on the Nucleolus and computed according to the algorithm of Aumann and Maschler appears to be a powerful approach. It requires that the agents reveal their true willingness to pay. It generates consistently superior results to HC and SA. NP is also more robust as the results vary far less. To achieve better results NP requires significantly fewer negotiation rounds which we show Sect. 4.4.

4.3   Amount of Revealed Information

According to the considerations in Sect. 3.4, the mediator requires $\frac{n^2+n}{2}$ queries to re-engineer the utility function of agent a. In only 4 out of 600 instances this number of queries was actually reached before the negotiations terminates. Two instances were from the set I_{25}^2 and two from I_{25}^{10}. However, it would have been possible for about a third of the small instances in I_{25}^2, I_{25}^5, and I_{25}^{10}, respectively, to re-engineer half the utility function. In contrast, re-engineering half the utility function was only possible for 1 out of 300 instances with 100 clauses.

The small instances with 25 clauses require 275 rounds but terminated on average after 150 rounds. The instances with 100 clauses require 4950 rounds but terminated on average after 1000 rounds. The number of agents seems to be of little influence. In conclusion, revealing information is less of a problem with an increasing number of negotiated clauses.

4.4   Convergence Behavior on a Selected Instance

The convergence behaviors of the three ANMs are shown by means of an example instance in Fig. 4. On the x-axis the negotiation rounds are shown ($r^{max} = 10,000$). The y-axis represents the social welfare, the solution found via CSA is defined as 1.0. Each point in Fig. 4 represents the social welfare of an updated and mutually accepted contract. During the negotiation of HC and NP, the agents accept only improving contracts. During SA the agents may accept inferior contracts stochastically. This explains the fluctuating social welfare of the current contract.

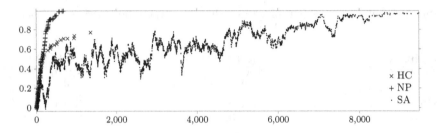

Fig. 4. Convergence of HC, SA, and NP for an I_{100}^2 instance. Each point represents new mutually accepted contract: $\times$ of HC, $\cdot$ of SA, and $+$ of NP.

Figure 4 shows that NP achieves superior solutions than SA with significantly less negotiation rounds. Therefore, NP appears not only more effective but also more efficient than SA. However, Fig. 4 represents only a single test instance. In contrast to HC and NP, SA almost constantly discovers new mutually accepted and improving contracts. Therefore, the question remains if it may be able to find even superior contracts to NP, if we allow for more negotiation rounds. An increase of r^{max} to say 20,000 rounds would increase the range of variation shown in Fig. 4 because the probability to accept inferior contracts during the early rounds of the negotiation process would significantly increase (see acceptance function of SA in Sect. 3.3).

5   Conclusion and Outlook

We have introduced a procedure for automated negotiations. The setting was introduced in [12]. We added compensation payments that are based on the Nucleolus and calculated in an efficient manner according to Aumann and Maschler [1]. The advantage of using this approach is a faster negotiation mechanism which identifies superior solutions. The disadvantage is that the agents have to reveal their willingness to pay and therefore disclose private information about the structure of their utility function. However, as computational results show, re-engineering utility functions might only become a problem for very small negotiation problems and not for more comprehensive ones.

Currently, we are working on a negotiation mechanism that combines the simulated annealing approach with Aumann-Maschler compensation payments. Additionally, different approaches to calculate compensation payments should be considered. It seems also necessary to analyze the influence of manipulated willingness to pay values. Finally, the presented NP should be applied to negotiation problems which are less stylized and take into account more structures from real-world problems.

Acknowledgement. The junior research group on *Computational Logistics* is funded by the University of Bremen in line with the Excellence Initiative of German federal and state governments.

References

1. Aumann, R.J., Maschler, M.: Game theoretic analysis of a bankruptcy problem from the talmud. J. Econ. Theory **36**(2), 195–213 (1985)
2. Buer, T., Haass, R.: Cooperative liner shipping network design by means of a combinatorial auction. Flex. Serv. Manuf. J. (2017). doi:10.1007/s10696-017-9284-8
3. Buer, T., Homberger, J., Gehring, H.: A collaborative ant colony metaheuristic for distributed multi-level uncapacitated lot-sizing. Int. J. Prod. Res. **51**(17), 5253–5270 (2013)
4. Dudek, G., Stadtler, H.: Negotiation-based collaborative planning in divergent two-tier supply chains. Int. J. Prod. Res. **45**(2), 465–484 (2007)
5. Eslikizi, S., Ziebuhr, M., Kopfer, H., Buer, T.: Shapley-based side payments and simulated annealing for distributed lot-sizing. In: Dolgui, A., Sasiadek, J., Zaremba, M. (eds.) 15th IFAC Symposium on Information Control Problems in Manufacturing (INCOM 2015), vol. 48, pp. 1592–1597. IFAC-PapersOnLine. Elsevier (2015)
6. Homberger, J.: Decentralized multi-level uncapacitated lot-sizing by automated negotiation. 4OR Q. J. Oper. Res. **8**, 155–180 (2010)
7. Homberger, J., Gehring, H., Buer, T.: Integrating side payments into collaborative planning for the distributed multi-level unconstrained lot sizing problem. In: 48th Annual Hawaii International Conference on System Sciences, pp. 1068–1077. IEEE Computer Society Press (2015)
8. Jacob, J., Buer, T.: Population-based negotiation of contract clauses and transportation request assignments. IFAC PapersOnLine **49**(12), 1862–1867 (2016). 8th IFAC Conference on Manufacturing Modelling, Management and Control MIM 2016, Troyes, France, 28–30 June 2016
9. Jennings, N., Faratin, P., Lomuscio, A., Parsons, S., Wooldridge, M., Sierra, C.: Automated negotiation: prospects, methods and challenges. Group Decis. Negot. **10**(2), 199–215 (2001)
10. Jonker, C.M., Robu, V., Treur, J.: An agent architecture for multi-attribute negotiation using incomplete preference information. Auton. Agents Multi Agent Syst. **15**(2), 221–252 (2007)
11. Kirkpatrick, S., Gelatt, C.D., Vecchi, M.P.: Optimization by simulated annealing. Science **220**, 671–680 (1983)
12. Klein, M., Faratin, P., Sayama, H., Bar-Yam, Y.: Negotiating complex contracts. Group Decis. Negot. **12**(2), 111–125 (2003)
13. Klein, M., Sayama, H., Faratin, P., Bar-Yam, Y.: The dynamics of collaborative design: insights from complex systems and negotiation research. Concurr. Eng. Res. Appl. **11**(3), 201–209 (2003)
14. Lopes, F., Wooldridge, M., Novais, A.Q.: Negotiation among autonomous computational agents: principles, analysis and challenges. Artif. Intell. Rev. **29**(1), 1–44 (2008)
15. Schmeidler, D.: The nucleolus of a characteristic function game. SIAM J. Appl. Math. **17**(6), 1163–1170 (1969)
16. Ziebuhr, M., Buer, T., Kopfer, H.: Agent-negotiation of lot-sizing contracts by simulated annealing with part-way resets. In: Klusch, M., Thimm, M., Paprzycki, M. (eds.) MATES 2013. LNCS (LNAI), vol. 8076, pp. 166–179. Springer, Heidelberg (2013). doi:10.1007/978-3-642-40776-5_16

Preference Modelling for Group Decision and Negotiation

University of Hohenheim

Stream Introduction: Preference Modelling for Group Decision and Negotiation

Adiel T. de Almeida[1] and Tomasz Wachowicz[2]

[1] Center for Decision Systems and Information Development, Universidade Federal de Pernambuco, Caixa Postal 7462, Recife 50630-970, Brazil
almeida@cdsid.org.br
[2] Department of Operations Research, University of Economics in Katowice, 1 Maja 50, 40-287 Katowice, Poland
tomasz.wachowicz@ue.katowice.pl

1 Overview

The bipolar nature of group decision and negotiation processes makes them difficult to analyze. On the one hand, a multitude of behavioral factors needs to be taken into consideration when examining the specificity of interaction between all stakeholders and parties involved. On the other, the formal issues related to the problem structuring and solving should be examined to assure that the compromising solution would be efficient and satisfying for all stakeholders. To provide a comprehensive support to the negotiating parties an integrated approach that takes into account both behavioral and formal issues should be applied. It requires combining the recent developments from such fields as sociology, psychology, economics, management science (decision making, operations research) and computer science. However, when the economic aspects of negotiation or group work are becoming more important to the stakeholders (such as the costs and profitability of the solutions implemented or their impact on the society) a bigger focus is put on improving the decision making mechanisms and developing the formal concepts that lead to best possible and efficient compromises.

To support the parties in negotiating the satisfying and efficient compromises the number of methods, algorithms and protocols may be implemented. They are designed and developed within a discipline called operation research (OR), and derive mainly from the methods of multiple criteria decision aiding (MCDA), voting and group decision making schemes, game theoretic notions of fair solution and fair division and others. The operations research offers also the set of approaches related to defining and structuring the negotiation and group decision problems developed within a soft operation research stream. Some behavioral or cognitive limitations in using the OR tools for negotiation support can also be taken into consideration within another stream called behavioral operations research. The whole arsenal of methods developed by OR is focused on a comprehensive analysis of the negotiation problem. It starts from precise definition of the problem under consideration (that takes into account the viewpoints of all stakeholders), defining the negotiation issues to be discussed, and the potential alternatives that comprise the proposals for the final negotiation contracts. Having the problem defined, the MCDA techniques are used to elicit the parties' individual preferences and build their scoring systems. These individual preferences seem to be crucial element of the whole support process, since they define real priorities, aspirations and reservations and comprise the feasible negotiation or group decision making space. The scoring systems built based on these preferences are used to evaluate precisely any offer or alternative for agreement that may appear during the negotiation or group work process. Having known the individual scoring systems, a mechanism may be developed to aggregate the individual decisions into a group one or find the alternative that satisfy the aspiration or reservation levels of all the parties. Here the game theoretic or voting approaches are most commonly used. In this way OR offers quite comprehensive approach to support the group decision makers and negotiators in their activities aimed at find profitable, efficient and economically justified solution.

In this chapter five papers are presented that were submitted to the stream on Preference Modeling for Group Decision and Negotiation (PMGDM) at GDN 2017 conference, and which are implementing OR procedures to support various problems of group decision making and negotiation. In the first paper entitled "The heuristics and biases in using the negotiation support systems" (Gregory E. Kersten, Ewa Roszkowska and Tomasz Wachowicz) an interesting problem of analyzing the cognitive biases in prenegotiation preparation is described. The authors examine the preference elicitation phase in the software supported negotiations, in which the negotiators build their individual negotiation offer scoring systems and try to identify the major errors and mistakes the negotiators make while using the tools for supporting their preference elicitation. They classify these errors as the results of the scaling biases revealed by the negotiators. Using the results of the online bilateral negotiation experiments conducted in Inspire system the authors show that the occurrence of scaling biases in electronic negotiation is quite common resulting in the scoring systems that do not reflect the negotiators preferences adequately and precisely. Yet, the authors emphasize that more detailed and advanced approach is required to identify what types of scaling biases, e.g. the omission, contraction or equalizing ones, appear most commonly during preference elicitation phase in negotiation support systems.

The next paper, entitled "Can the holistic preference elicitation be used to determine accurate negotiation offer scoring systems? A comparison of direct rating and UTASTAR technique" (Ewa Roszkowska, Tomasz Wachowicz) discovers the nuances related to different techniques of preference elicitation in the principal-agent context. The authors consider two different setups of preference elicitation protocol that can be offered to the negotiators in the negotiation support system, both based on additive preference model. In the first setup the preferences are defined by the negotiators themselves using the direct rating assignment, while in the second one the disaggregation paradigm is applied, which allows negotiators to define their preferences by means of ordering the offers examples. It is assumed, that the latter one is less cognitively demanding, since no quantitative preferential information needs to be provided and the negotiators compare the two offers only at a time considering which is better or whether they are equivalent. The scoring systems resulting from both elicitation approaches are compared with respect to their accuracy to the principal preference system. Surprisingly, using an ample dataset of electronic negotiation experiments, the authors find that the second approach, which operates with more general preferential information, does not lead to the scoring system of lesser accuracy that the one based on direct ratings.

In the paper "Choosing a Voting Procedure for the GDSS GRUS" (Rachel Perez Palha, Pascale Zarate, Adiel Teixeira de Almeida, and Hannu Nurmi) deals with the question of "who should decide the voting method?" A framework for choice of a voting procedure in a business decision context is applied in order to to choose which voting procedure best suits the environment of the Group Decision Support System GRoUp Support (GRUS). Voting rules may be applied in this type of system.

The paper "A group decision outranking approach for the agricultural technology packages selection problem" (Pavel A. Álvarez Carrillo, Juan C. Leyva López, Omar Ahumada Valenzuela) deals with the selection problem of technological packages identifying criteria and alternatives in a group decision. The proposed model is based on outranking methods, considering a great divergence among the DMs. It can be used for defining rural credits for technological packages in order to improve competitiveness and profitability of farmers.

Finally, Francis Marleau Donais, Irène Abi-Zeid and Roxane Lavoie in their paper entitled "Building a shared model for multi-criteria group decision making: Experience from a case study for sustainable transportation planning in Quebec City" present an interesting insight into the process of building a common model of the real world multi-party decision making problem. They consider an issue of better-integrated sustainable transportation for street rehabilitation in Quebec City, Canada. For this purpose, they proposed the comprehensive protocol for problem structuring, design and support. First, they recommend the setup for structuring phase that consists of preparatory meetings and group workshops. Then, they move to preference elicitation phase that consists of two sub-phases: building the evaluation scales for criteria and determining the a common system of values. Here the group workshops as well as sub-group workshops are proposed and, from the technical viewpoint – MACBETH algorithm is applied. Finally validation phase is implemented to check the reliability of the preference model built. While the model is validated, it is used then to alternative selection. Apart from describing the technical issues related to building such a decision support protocol the authors describe also the results they obtain after it was applied to the problem of evaluating the Quebec City streets. They authors also comment on the pragmatic and organizational problems they faced while implementing their approach with a group of Quebecois professionals involved in the project.

We thank our reviewers and advisors for their valuable comments and suggestions that helped the authors to improve their papers and make the PMGDM stream discussions fruitful and inspiring.

Choosing a Voting Procedure for the GDSS GRUS

Rachel Perez Palha[1]([✉]), Pascale Zarate[2], Adiel Teixeira de Almeida[1], and Hannu Nurmi[3]

[1] CDSID - Center for Decision Systems and Information Development,
Universidade Federal de Pernambuco, Caixa Postal 7462, Recife, Pernambuco 50630-970, Brazil
{rachelpalha,almeida}@cdsid.org.br
[2] IRIT, Toulouse University, Toulouse, France
pascale.Zarate@ut-capitole.fr
[3] University of Turku, Turku, Finland
hnurmi@utu.fi

Abstract. In group decision-making, the use of Group Decision Support Systems is increasing and in some groups, a facilitator is required to improve communication among participants. The facilitator has several roles in this situation, which include helping decision makers (DMs) to decide which type of aggregation they would prefer in each decision context. Whenever DMs have different objectives regarding the same problem, they might decide a consensual decision is no longer possible. Therefore, other types of aggregation are required. Voting rules are strongly applied in this type of situation. However, the question that arises is: who should decide the voting method? In this article, a framework for choice of a voting procedure in a business decision context is used. It takes the facilitator's preferences into account while it seeks to choose which voting procedure best suits the environment of the Group Decision Support System GRoUp Support (GRUS).

Keywords: GDSS · Facilitator · MCDM voting choice · Choice of a voting procedure · Preference analysis

1 Introduction

Dufner *et al.* [1] state that most managers spend between 25% and 80% of their productive time in meetings trying to solve problems and make decisions focusing on shared objectives. They also verified that approximately 50% of this time is wasted due to information being lost and distorted, and because decisions made are suboptimal. They also state that it is a common belief that using Group Decision Support Systems (GDSS) may reduce these losses and increase the productivity of a group. Therefore, several GDSS have been proposed in the literature.

Colson [2] presented a modification to ARGOS, by adding a new feature: JUDGES. His objective was to provide a GDSS to help a jury choose the best student's final work by the Belgian Operations Research Society based on some criteria decided by the jury. In the first part of the model, the jury uses ARGOS to individually evaluate each alternative by using ELECTRE [3] or PROMETHEE [4]. In the second part, JUDGES presents the jury with the aggregated results based on different methods: voting rules or consensus. Damart *et al.* [5] presented IRIS, which is based on ELECTRE-TRI and combines the disaggregation approach and preference elicitation. The profiles are defined *a priori,* and the weights and cutting level are defined by applying the disaggregation approach. The facilitator's role in this context is to help the DMs reach a consensus. FlowSort-GDSS [6] was proposed in the risk and reliability context but does not make use of a facilitator.

Problems in coordinating groups who use GDSS may arise, such as fragmentation into subgroups and overall confusion of group members as to who is responsible for what by when [1]. Colson [2] found that studies on GDSS are more concerned about facilitating group meetings than with the multicriteria framework itself. When a group has sub-divided into distributed groups, a challenge which the group needs to overcome is how best to tackle asynchronous mediated communication. Thus, Kim *et al.* [7] conduct a study to examine the effect of system restrictiveness under this type of environment.

What the role of the facilitator should be is a topic that has also been studied over the years [8]. Ackermann [8] interviewed group members in order to find out what their perception of the facilitator's role is and to address some practical suggestions from professionals working in this field. Other frameworks seek to help facilitators to focus on important issues regarding the use of GDSS and helping the group to interact and to build a model which reflects their shared objectives. See, for example, those proposed by Ebadi *et al.* [9] and Adla *et al.* [10]. Also, some electronic agents have been presented in different GDSS such as in Rigopoulos *et al.* [11] and Jahng and Zahedi [12].

There are GDSS where the group's intention is to reach a consensual decision, which prompts the facilitator to play a role where he/she has to help the group to discover their shared objectives, the available alternatives, and a compromise solution. In other contexts, the group does not wish to reach consensus, but some interaction is necessary. Therefore, group decision problems can be divided into two main streams: (1) the DMs share the same objectives concerning a problem and (2) the DMs have different objectives regarding a problem. In the latter, the DMs want to have their point of view considered in the analysis but are prepared to accept a decision different from their own in favor of the group's [13]. It is in this context that voting rules are frequently used to reach a single compromise solution.

Several voting rules have been presented over the years. They have different features and are applicable in different contexts. All authors who have presented such rules were trying to avoid different voting paradoxes [14]. These paradoxes may allow the analyst or one of the DMs, who plays the role of an analyst, to manipulate the system to their own personal advantage. Thus, de Almeida and Nurmi [15] proposed a framework to deal with the decision process related to choosing a voting procedure for a business organization. The idea of using such a framework is to reinforce the commitment of the DMs with the solution since all of them had the opportunity to get involved in the choice of the method. Some concern on how to choose multicriteria decision aiding (MCDA) methods have been tackled in different contexts. Gillian *et al.* [16] emphasized that several MCDA methods are available and they are not compatible with every decision context. Therefore, the choice of the wrong method may drive the solution to be misleading or unsatisfactory, causing useful techniques to be judged inappropriate and losses in energy, time and money due to wrong decisions. The choice of the method can be analyzed as a multicriteria problem as well, and this choice depends on the type of problem, goals of the DMs and desired properties of the compromise solution [17]. De Almeida *et al.* [13] presented a twelve-steps procedure to guide the analyst in the choice of a method compatible with the problem faced, and the DM's rationality and objectives. In addition, Roy and Słowinski [18] formulated some questions to help the analyst in the choice of the MCDA method more compatible with the decision context faced.

Furthermore, it should be noticed the ethical issue behind this question. The analyst's preference on the method choice may have ethical considerations [13]. Rauschmayer et al. [19] brings such an ethical considerations regarding to the modeling process, particularly for the choice of the method and its parameterization. It must be taken into account that distortions in the results cannot be made for interests other than the DMs. Also, the assumptions of the model must be shared with the DMs.

In this article, the framework proposed by de Almeida and Nurmi [15] is used to choose a voting rule, compatible with a facilitator's preferences, that applies a GDSS to conduct a synchronous group decision process. The facilitator, in such a case, is not familiar with all voting rules. Thus, the choice of the more appropriate voting rule is not intuitive, and the use of the framework is justified. Since the DMs are not directly involved in this choice, it is more difficult for any one of them to introduce bias into the process and this framework might drive the whole process to a Social Choice compatible with the group of DMs.

The article is divided into four sections. In Sect. 2 the context of the problem is presented, while Sect. 3 presents the experimental application. Section 4 draws conclusions and indicates lines of future research that could be usefully developed.

2   Context of the Problem

A Group-Decision Support System (GDSS) built on a web-based platform, called GRoUp Support System (GRUS) [20], is modularized in order to allow a facilitator to build the best structure for the problem that is being analyzed. In this system, there are two types of users: decision makers (DMs) and a facilitator, who is responsible both for the protocol of the group decision process and for leading the interaction process. The DMs have to put forward their ideas as to the electronic interaction in the first step, called "brainstorming", where they suggest, be it anonymously or not, the criteria they believe to be related to the problem, and also alternatives for solving the problem.

Once the first step is over, the facilitator leads the group to a verbal interaction, where they have to cluster the criteria and alternatives. They finally evaluate the alternatives regarding these criteria. For these evaluations, a suitability equation function is required, which calculates the score of each alternative using Choquet Integral. This function is a kind of preference criterion. Another calculation is made with the Simple Additive Weight (SAW) [21]. To reach a final decision, the facilitator has to lead a consensus process, which is usually time-consuming and wearing, in addition to which it is usual to face situations where the DMs have different objectives concerning the same problem.

Currently, the process implemented considers that the group members' objectives regarding the problem are the same. This is the first type of group decision problem presented by de Almeida *et al.* [13]. Therefore, the process was built to reach a consensual decision that has to be achieved in a face-to-face group meeting and will require DMs to change their positions until a potential commitment is found. The question that is addressed here is how to lead the process whenever the group members have different objectives for the problem.

When the DMs do not share the same objectives, even if they agree to evaluate the alternatives using the same criteria, they have different perceptions about the meaning of each criterion. In these cases, the DMs propose their individual ranking of the alternatives, and they are aggregated based on DMs' final choices. One way of running this type of aggregation is by applying a voting procedure. Several voting procedures have been proposed over the years. Nurmi [22] presented a comparative analysis and showed that each of the methods is associated with serious drawbacks.

Usually, an analyst is responsible for choosing a voting procedure that is compatible with the needs of the group to reach a group decision. The choice of the best voting rule has been discussed over the years in the literature [23] and how to define setting out to do this usually relies on the properties of each method. Thus, some articles have compared voting rules by considering some aspects related to the properties sought for in these methods [24–28]. The main problem is that it is usually the analyst who chooses the voting rule but he/she is not supposed to be the best person to make this choice since he/she will not deal with the consequences of the social choice, as discussed by de Almeida and Nurmi [15].

To allow the DMs to have their preferences considered in this analysis, de Almeida and Nurmi [15] propose a framework to aid the choice of a voting procedure for decisions in a business decision context. Nurmi [23] conducted a numerical application of this method by considering some of the voting properties, which are characterized by the capability of a rule to overcome a voting paradoxes, as criteria for analyzing the voting procedures, which played the role of alternatives. The main idea is to consider a decision matrix where the voting procedures are the alternatives that are evaluated by considering some criteria. These criteria are divided into two main streams: voting properties and criteria related to the context of the problem. The latter is associated with how easily they can be applied. The decision matrix is evaluated by using a multiple-criteria decision method, which is selected by considering the characteristics of the methods and the problem itself and guided by the procedure proposed by de Almeida *et al.* [13].

Since in the context of GRUS, the DMs do not undertake the decision process without a facilitator, and the group depends on the facilitator right from the very beginning of the process, it is important to apply the framework, thereby allowing the facilitator to decide which voting procedure would be best suited for this application. The application of the framework avoids manipulation on behalf of one or more parties, even when it is applied considering the facilitator's preferences. The analysis of the ease of voting depending on the procedure was not considered, because the ranking of alternatives is delivered by the application itself, and does not require more cognitive effort from the DMs. Therefore, only the voting properties were considered in the analysis.

3   Experimental Application

In this study, one of the authors plays the role of the analyst or facilitator. She was interviewed to express her preferences regarding the selection of the voting procedure to be used to aggregate the group members' preferences in one experiment. A subset of methods was considered in this analysis. The voting rules considered were: Amendment [14], Copeland [22], Dodgson [22], Maxmin, Kemeny [29], Plurality [22], Borda [30], Approval Voting [31], Black [22], Plurality runoff [22], Nanson [32] and Hare [22]. Other methods might be available but were not considered in this analysis such as the quartiles method [33, 34] and those that consider partial information [35, 36].

In order to evaluate all alternatives, namely voting rules, voting properties were considered. The criteria used to evaluate the voting procedures were as follows:

(a) the procedure should always choose a Condorcet winner when there is one. A Condorcet winner is the alternative which defeats all alternatives in pairwise comparisons [14];

(b) the procedure should never choose a Condorcet loser when there is one. The Condorcet loser is the opposite of the Condorcet winner. Thus, it is an alternative that is defeated by all other alternatives in pairwise comparisons [14];

(c) the procedure makes use of the strong Condorcet criterion, which is satisfied by all systems that always end up with a strong Condorcet winner when there is one. A strong Condorcet winner is an alternative that is ranked first by all individuals [14];

(d) the procedure makes use of monotonicity. This can be expressed as "if an alternative x wins in a given profile P when a certain procedure is being applied, it should also win in the profile P' obtained from P by placing x higher in some individuals' preference rankings, *ceteris paribus*." [14]. This means that additional support cannot transform a winning alternative into a non-winning alternative;

(e) The Pareto criterion exists whenever all voters strictly prefer x to y, and thus y cannot be elected [14];

(f) the procedure presents Consistency that is satisfied by those systems that have the following property. Suppose that the group is split into two groups so that the same alternative is chosen in both groups. Then the procedure is consistent if the same alternative is chosen if the procedure is applied to the group as a whole [14];

(g) the procedure presents the Chernoff property, which means that if an alternative is a winner in a set of alternatives, it has to be the winner in every subset of these alternatives [14];

(h) the procedure is consistent with the property of independence of irrelevant alternatives. A procedure presents this property if two profiles have identical rankings over a pair of alternatives. Thus, the collective ranking over this pair is the same in these two profiles, regardless of the rankings over the other pairs [37]; and

(i) the procedure presents the invulnerability of the no-show paradox, which is a condition in which an elector may achieve a better result by not voting, thus prompting him/her to manipulate the voting result by abstaining [14].

Several authors advocate that these characteristics are binary so that a procedure only may have one out of two conditions: either it has the property, or it does not [15].

Whenever the procedure has the property sought, it will be represented by 1 (one), and when it does not, the representation is 0 (zero). Table 1 presents the evaluation of the 12 voting procedures considered, which were evaluated by considering that the criteria of evaluation ought to be binary. It was not taken into account any criterion related to the context of the problem because any of the voting rules would receive the same information as input and would give the same type of output to the DMs. Furthermore, the hardness to implement the voting rule into the GRUS system could have been considered, but it was not evaluated the difficulties related to creating the algorithm inside the system. Therefore, only the voting properties were considered as criteria.

Table 1. Voting procedures vs. voting procedures

Voting rule	Criteria								
	a	b	c	d	e	f	g	h	i
Amendment	1	1	1	1	0	0	0	0	0
Copeland	1	1	1	1	1	0	0	0	0
Dodgson	1	0	1	0	1	0	0	0	0
Maximin	1	0	1	1	1	0	0	0	0
Kemeny	1	1	1	1	1	0	0	0	0
Plurality	0	0	1	1	1	1	0	0	1
Borda	0	1	0	1	1	1	0	0	1
Approval voting	0	0	0	1	0	1	1	0	1
Black	1	1	1	1	1	0	0	0	0
Plurality runoff	0	1	1	0	1	0	0	0	0
Nanson	1	1	1	0	1	0	0	0	0
Hare	0	1	1	0	1	0	0	0	0

During the interview, the facilitator was seen to have a non-compensatory rationality regarding the group decision processes and, therefore, a non-compensatory method was selected to evaluate the set of alternatives. The preferences concerning the criteria were provided in order of importance to allow the weights to be calculated. The evaluation was made by considering a five-level scale as presented in Table 2 since it was not necessary to provide a complete order.

Table 2. Voting procedures notation scale

VU	Very unimportant	Which means that in this context the criteria do not bring any important feature to the problem
NI	Not important	Which means that in this context the criteria do not bring more than two important features to the problem
SS	So-so	Which means that in this context the facilitator is indifferent to the features brought by the criteria
I	Important	Which means that in this context the criteria bring at least one important feature of the problem
VI	Very important	Which means that in this context the criteria bring more than two very important features to the problem

The ordinal values were converted to a numeric scale, where VU represented 0.2 on a scale from 0 to 1 and VI represented 1. This parametrization was taken into account because a value of 0 meant that the criteria had no relevance at all for the facilitator and it would not be considered in the analysis. Since it does not make sense to consider irrelevant criteria, the least valuable criteria had to be assigned a value of 0.2. The numerical scale considered was VU = 0.2, NI = 0.4, SS = 0.6, I = 0.8, and VI = 1, which are related to the levels presented in Table 2. The facilitator used this scale to evaluate each criterion and, once this step was ended, the values were normalized by considering the scaling process presented in Eq. 1.

$$\pi'_i = \frac{\pi_i}{\sum_j \pi_j} \tag{1}$$

Where: π'_i is the scaled weight value of criterion i.

π_i is the weight value of criterion i on the five point scale.

$\sum_j \pi_j$ is the sum of the weights of all criteria.

The preferences expressed by the facilitator were related to each of the positions on the scale. Criteria c and e were considered "very important", the former because it is imperative that the solution is in the set of non-dominated alternatives and the later because it is mandatory that the solution is Pareto-optimal. Criteria a, b and d were considered "important": the first and the second because a procedure has to be reliable and guarantee such that the best alternative in pairwise comparison will be the winner and the worst will not, if those alternatives do exist, and the third one because additional support should not lead a winning alternative to become non-winning. No criterion was evaluated as "so-so". The "not important" criteria were considered to be f and i, the former because the analysis will hardly ever be made considering subsets of DMs and the later because the DMs will not be able to manipulate the analysis at this point. Finally, the "very unimportant" criteria were considered to be g because it is unlikely that the group will decide to visualize a subset of alternatives during the analysis. At last, criteria h was excluded from the analysis because all procedures fail on it. It is said to be important to have a procedure that is independent of irrelevant alternatives, but none of the above are, which leads to these criteria being excluded from the analysis. As an example of the procedure to normalize the weights, let us consider criterion "a", which has a nominal weight of $\pi_a = 0.80$. The sum of all criteria is $\sum_j \pi_j = 5.4$. Thus, $\pi'_a = \pi_a / \sum_j \pi_j = 0.8/5.4 = 0.148$. The same calculations were used in all criteria and the results are presented in Table 3.

Table 3. Weights of criteria

Type of weight	Criteria							
	a	b	c	d	e	f	g	i
Scaled weights	0.148	0.148	0.185	0.148	0.185	0.074	0.038	0.074
Weights	0.80	0.80	1.00	0.80	1.00	0.40	0.20	0.40

The analysis was conducted by applying two outranking procedures: ELECTRE III [3] and PROMETHEE I [4]. Two different procedures were used to allow the results from both to be compared and to verify if by changing the outranking method, this would lead the analysis to a different result.

In the procedure using ELECTRE III, three concordance indices [38] were considered, namely: $s_1 = 0.9$, $s_2 = 0.85$ and $s_3 = 0.8$. The main idea of evaluating the data by applying different concordance indices was to verify if the kernel [3] would be modified by increasing the strength of the concordance coalition. No considerations of discordance indices were made because the values were only binary, and thus, all the differences between evaluations are 0 or 1, which enables this index to be used in this environment. In all cases, three kernels were found: (1) Copeland; (2) Black; and (3) Kemeny. The Hasse Diagram [39] presented in Fig. 1 enables the relationship between all alternatives to be visualized.

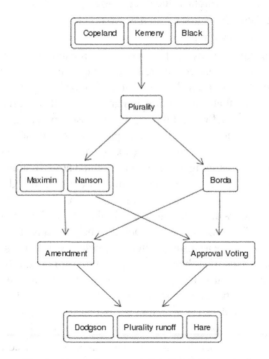

Fig. 1. Hasse diagram of ELECTRE III method

When PROMETHEE I was used, the Hasse Diagram presented in Fig. 2 was found. The results are similar, in the sense that those based on Copeland, Kemeny, and Black were seen, once again, to have insignificant differences. This outcome is expected since the evaluation in all their criteria was the same, and the synthesis of their results was very similar. In this method, the difference was that the Borda rule was incomparable to these three rules. Thus, its position changed from third to first. In addition, when ELECTRE III method was used, the Plurality rule remained in second place, but instead of outranking Maximin and Nanson, it became incomparable. The other rules, even after undergoing some modifications as to their preference relations would not be chosen to solve the problem. The results were also compared with PROMETHEE II [4], and Copeland, Kemeny, and Black remained in the same place, but the Borda rule dropped to third.

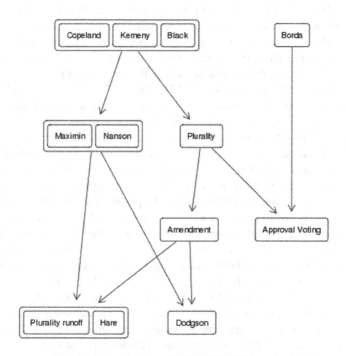

Fig. 2. Hasse diagram of PROMETHEE I method

The three methods used presented Copeland, Kemeny and Black procedures as a tie, and PROMETHEE I also presented Borda's rule as not being comparable with these voting procedures. This behavior moves the problem on to another question which is how to choose the method when such a situation arises. This indifference might have occurred due to missing criteria, thus, by considering other criteria, one could tie up the procedures. It could take into account other voting properties, the ease with which the method inside GRUS can be implemented and the possibility of adapting the procedures to a partial information environment.

The results can be easily interpreted, and the next step after choosing one of these voting rules is to create a module to work inside GRUS in order to aggregate the DMs' preference information whenever they are not interested in reaching a consensual decision. The three methods that were presented in the first position in all methods are all distance-based methods but have different types of input information and methodologies that provide DMs with a final ranking of alternatives.

4 Conclusion

This article has presented an application of the framework proposed by de Almeida and Nurmi [15] for choosing a voting procedure in the business context to decide which voting rule is best suited for aiding a facilitator using a GDSS called GRUS when the group does not wish to reach a consensual decision. During the application, some difficulties arose. One of them is how to go about deciding on which of the three methods to choose. Maybe the inclusion of other criteria, such as the ease of applying the rule in the decision context, may present a different solution to this problem.

In their article, de Almeida and Nurmi [15] presented their framework and suggested that the analysis of the voting procedures could be made on a five-level scale if it was decided to include a subjective criterion in the evaluation. When Nurmi [23] presented his application of this framework, he considered that all voting procedures evaluated under the voting properties only had a binary evaluation when he considered whether or not a voting procedure has a certain property. Anyhow, de Almeida and Nurmi [15] also showed that maybe this evaluation could be less strict by considering the proportion of cases where the voting procedures actually present the property. Lepelley and Valognes [26] proposed to verify the efficiency in Kim and Roush's voting procedure [28] by applying an Impartial Anonymous Culture condition (IAC) by calculating the probability of a voting situation to verify if the procedure was efficient for two voting properties: the Condorcet winner and the Condorcet loser. Fishburn and Gehrlein [25] presented a simulation to verify the efficiency of a simple majority of some voting procedures. Nurmi [27] compared distance-based voting rules in ranking environments. Therefore, in future research, it is important to run a simulation to verify the proportion in which each of these properties actually occurs in each method and to compare these with the binary results and find out if the voting rule would change in this scenario.

The choice of the method is not the only challenge the facilitator has to face in this type of aggregation. He/she must also build an agenda with the participants to decide which weight will be assigned to each of the DMs. Also, it is important to make sure the DMs want to empower the facilitator in this decision. Finally, it is important remark that the analyst must not impose their preferences in the model. This may arise ethical issue behind, particularly for the choice of the method and its parameterization.

Acknowledgment. The authors are grateful to the Brazilian Research Council (CNPq) for their financial support of the research contained in this paper.

References

1. Dufner, D., Hiltz, S.R., Johnson, K., Czech, R.: Distributed group support: the effects of voting tools on group perceptions of media richness. Group Decis. Negot. **4**, 235–250 (1995). doi: 10.1007/BF01384690
2. Colson, G.: The OR's prize winner and the software ARGOS: how a multijudge and multicriteria ranking GDSS helps a jury to attribute a scientific award. Comput. Oper. Res. **27**, 741–755 (2000). doi:10.1016/S0305-0548(99)00116-1

3. Roy, B., Bouyssou, D.: Aide multicritère à la decision: méthodes et cas. Economica, Paris (1993)
4. Brans, J.P., Vincke, P., Mareschal, B.: How to select and how to rank projects: the promethee method. Eur. J. Oper. Res. **24**, 228–238 (1986). doi:10.1016/0377-2217(86)90044-5
5. Damart, S., Dias, L.C., Mousseau, V.: Supporting groups in sorting decisions: methodology and use of a multi-criteria aggregation/disaggregation DSS. Decis. Supp. Syst. **43**, 1464–1475 (2007). doi:10.1016/j.dss.2006.06.002
6. Lolli, F., Ishizaka, A., Gamberini, R., Rimini, B., Messori, M.: FlowSort-GDSS – a novel group multi-criteria decision support system for sorting problems with application to FMEA. Expert Syst. Appl. **42**, 6342–6349 (2015)
7. Kim, Y., Hiltz, S.R., Turoff, M.: Coordination structures and system restrictiveness in distributed group support systems. Group Decis. Negot. **11**, 379–404 (2002). doi:10.1023/A:1020492305910
8. Ackermann, F.: Participants' perceptions on the role of facilitators using group decision support systems. Group Decis. Negot. **5**, 93–112 (1996). doi:10.1007/BF02404178
9. Ebadi, T., Purvis, M., Purvis, M.: A distributed and concurrent framework for facilitating cooperation in dynamic environments. In: Proceedings - 2010 IEEE/WIC/ACM International Conference on Intelligent Agent Technology, pp. 287–294, IAT (2010). doi:10.1109/WI-IAT.2010.260
10. Adla, A., Zarate, P., Soubie, J.L.: A proposal of toolkit for GDSS facilitators. Group Decis. Negot. **20**, 57–77 (2011). doi:10.1007/s10726-010-9204-8
11. Rigopoulos, G., Karadimas, N.V., Orsoni, A.: Facilitating group decisions through multicriteria analysis and agent based modeling. In: First Asia International Conference on Modelling & Simulation (AMS 2007), pp. 533–538 (2007). doi:10.1109/AMS.2007.40
12. Jahng, J., Zahedi, F.: Intelligent electronic facilitator: increasing GDSS effectiveness and making web-based GDSS possible. In: AMCIS 1998, Paper 161 (1998)
13. de Almeida, A.T., Cavalcante, C.A.V., Alencar, M.H., Ferreira, R.J.P., de Almeida-Filho, A.T., Garcez, T.V.: Multicriteria and Multiobjective Models for Risk, Reliability and Maintenance Decision Analysis, vol. 231. Springer (2015). doi:10.1007/978-3-319-17969-8
14. Nurmi, H.: Voting Paradoxes and How to Deal with Them. Springer, New York (1999). doi:10.1007/978-3-662-03782-9
15. de Almeida, A.T., Nurmi, H.: A framework for aiding the choice of a voting procedure in a business decision context. In: Kamiński, B., Kersten, G.E., Szapiro, T. (eds.) GDN 2015. LNBIP, vol. 218, pp. 211–225. Springer, Cham (2015). doi:10.1007/978-3-319-19515-5_17
16. Gilliams, S., Raymaekers, D., Muys, B., Orshoven, J.V.: Comparing multiple criteria decision methods to extend a geographical information system on afforestation. Comput. Electron. Agric. **49**, 142–158 (2005). doi:10.1016/j.compag.2005.02.011
17. Górecka, D.: On the choice of method in multicriteria decision aiding process. Multiple Criteria Decis. Making **6**, 81–108 (2011)
18. Roy, B., Słowiński, R.: Questions guiding the choice of a multicriteria decision aiding method. EURO J. Decis. Process. **1**, 69–97 (2013). doi:10.1007/s40070-013-0004-7
19. Rauschmayer, F., Kavathatzopoulos, I., Kunsch, P.L., Le Menestrel, M.: Why good practice of OR is not enough-ethical challenges for the OR practitioner. Omega **37**, 1089–1099 (2009). doi:10.1016/j.omega.2008.12.005
20. Zaraté, P., Kilgour, D.M., Hipel, K.: Private or common criteria in a multi-criteria group decision support system: an experiment. In: Yuizono, T., Ogata, H., Hoppe, U., Vassileva, J. (eds.) CRIWG 2016. LNCS, vol. 9848, pp. 1–12. Springer, Cham (2016). doi:10.1007/978-3-319-44799-5_1

21. Keeney, R.L., Raiffa, H.: Decision with Multiple Objectives: Preferences and Value Trade-Offs. Wiley, New York (1976)
22. Nurmi, H.: Comparing Voting Systems. D. Reidel Publishing Company (1987). doi: 10.1007/978-94-009-3985-1
23. Nurmi, H.: The choice of voting rules based on preferences over criteria. In: Kamiński, B., Kersten, G.E., Szapiro, T. (eds.) GDN 2015. LNBIP, vol. 218, pp. 241–252. Springer, Cham (2015). doi:10.1007/978-3-319-19515-5_19
24. Nurmi, H.: Voting procedures: a summary analysis. Br. J. Polit. Sci. **13**, 181–208 (1983). doi: 10.1017/S0007123400003215
25. Fishburn, P.C., Gehrlein, W.W.: Majority efficiencies for simple voting procedures: summary and interpretation. Theory Decis. **14**, 141–153 (1982). doi:10.1007/BF00133974
26. Lepelley, D., Valognes, F.: On the Kim and roush voting procedure. Group Decis. Negot. **8**, 109–123 (1999)
27. Nurmi, H.: A comparison of some distance-based choice rules in ranking environments. Theory Decis. **57**, 5–24 (2004). doi:10.1007/s11238-004-3671-9
28. Kim, K.H., Roush, F.W.: Statistical manipulability of social choice functions. Group Decis. Negot. **5**, 263–282 (1996)
29. Saari, D.G., Merlin, V.R.: A geometric examination of Kemeny's rule. Soc. Choice Welfare **17**, 403–438 (2000). doi:10.1007/s003550050171
30. Borda, J.C.: Mémoire sur les élections au scrutin. Histoire de l'Académie Royale Des Sciences (1781)
31. Brams, S.J., Fishburn, P.C.: Approval voting. Am. Polit. Sci. Rev. **72**, 831–847 (1978)
32. Nanson, E.J.: Methods of election. Trans. Proc. R. Soc. Victoria Art **XIX**, 197–240 (1883)
33. Morais, D.C., de Almeida, A.T.: Group decision making on water resources based on analysis of individual rankings. Omega **40**, 42–52 (2012). doi:10.1016/j.omega.2011.03.005
34. de Almeida-Filho, A.T., Monte, M.B.S., Morais, D.C.: A voting approach applied to preventive maintenance management of a water supply system. Group Decis. Negot. **26**(3), 523–546 (2017). doi:10.1007/s10726-016-9512-8
35. Cullinan, J., Hsiao, S.K., Polett, D.: A borda count for partially ordered ballots. Soc. Choice Welfare **42**, 913–926 (2014). doi:10.1007/s00355-013-0751-1
36. Ackerman, M., Choi, S.Y., Coughlin, P., Gottlieb, E., Wood, J.: Elections with partially ordered preferences. Publ. Choice **157**, 145–168 (2013). doi:10.1007/s11127-012-9930-3
37. Arrow, K.J.: Social Choice and Individual Values, 2nd edn. Wiley, New York (1963)
38. Figueira, J., Mousseau, V., Roy, B.: Electre methods. In: Figueira, J., Greco, S., Ehrogott, M. (eds.) Multiple Criteria Decision Analysis: State of the Art Surveys. International Series in Operations Research & Management Science, vol 78. Springer, New York (2005)
39. Skiena, S.: Implementing Discrete Mathematics: Combinatorics and Graph Theory with Mathematica. Addison-Wesley, Redwood City (1990)

Building a Shared Model for Multi-criteria Group Decision Making

Experience from a Case Study for Sustainable Transportation Planning in Quebec City

Francis Marleau Donais[1(✉)], Irène Abi-Zeid[2], and Roxane Lavoie[1]

[1] Graduate School of Land Management and Regional Planning,
Université Laval, Québec, Canada
francis.marleau-donais.1@ulaval.ca, roxane.lavoie@esad.ulaval.ca
[2] Department of Operations and Decision Systems, Université Laval, Québec, Canada
irene.abi-zeid@osd.ulaval.ca

Abstract. Shared procedures to build a consensus within a group decision process are sometimes used in multi-criteria decision-making. Facilitators often face several challenges and the solutions to overcome them are scarce and not well documented. This paper presents a case study within a decision framework that combines problem structuring with the multi-criteria decision aid method MACBETH in order to build a shared preference model. The framework was applied in a transportation planning context with a group of professionals from Quebec City, Canada to assess and rank streets as a function of their potential to become Complete Streets. The analysis of the process showed that difficulties in expressing preferences, access to data during workshops, group size, group discussion management, and project length were encountered. Nonetheless, the proposed framework and the use of sub-groups to build criteria scales were a way to overcome these challenges and allowed us to successfully complete the project.

Keywords: Group decision-making · MACBETH · Problem structuring · Preference modelling · MCDM · Case study

1 Introduction

Over the last decades, several methods for multi-criteria decision-making (MCDM) and problem structuring have been developed to model decision makers preferences and to facilitate group decisions in organizations. Problem structuring has been widely accepted as an integral part of the MCDM process. Problem structuring methods (PSM), also called Soft OR, and Value-Focused Thinking (VFT) are two major approaches used to structure decision problems [1, 2]. More recently, various frameworks that combine problem structuring and MCDM have been theorized, conceptualized, and applied [2, 3].

Decision makers in a group setting often have differing points of view for various reasons including conflicts, misunderstandings and uncertainties [4]. To deal with these differences, MCDM group preference modelling usually implies one of three

© Springer International Publishing AG 2017
M. Schoop and D.M. Kilgour (Eds.): GDN 2017, LNBIP 293, pp. 175–186, 2017.
DOI: 10.1007/978-3-319-63546-0_13

elementary coping procedures: sharing preferences where the group is assumed to behave as a single decision maker, aggregating individual preferences, or comparing individual preferences as a basis for discussion [4]. The goal in sharing preferences procedures is to obtain a common vision by consensus, using group workshops (GWS), where differences are discussed and agreements are negotiated. However, sharing procedures are time-consuming and require that all participants are present in every GWS [4]. Furthermore, sharing MCDM GWS are quite challenging and demanding for facilitators who must have a high level of expertise and be aware of the various underlying group dynamics and interactions. Although several methods have been developed to facilitate the inclusion of multiple stakeholders by aggregating individual preferences [5–8], the literature on how to develop expertise as a facilitator in a sharing group decision procedure and how to overcome the difficulties encountered are scarce and the solutions not well documented.

This paper describes a framework combining problem structuring and MCDM using different types of workshops to reach consensus among a group of eleven stakeholders with varying technical and managerial backgrounds. A shared group decision framework is proposed by integrating subgroup workshops (SWS) to the usual MCDM GWS. Organizing participation by breaking up groups in sub-groups is a method sometimes used to facilitate group decision-making in PSM [9].

We applied the proposed framework to a multi-criteria decision process to better integrate sustainable transportation for street rehabilitation in Quebec City, Canada. The recent evolution of transportation planning to include a more sustainable approach had a considerable impact on transportation projects and related decision-making processes. Integration of multidimensional aspects and inclusion of stakeholders with different expertise and interests have made decision-making and transportation project assessments more complex [10]. To implement sustainable and active transportation in North America, one of the biggest movements in the last decades was, and still is, the Complete Streets movements. This movement encourages elected officials to adopt policies, and planners to design streets that are safe, accessible and comfortable for everyone, regardless of their transportation mode or physical condition [11]. However, to our knowledge, there is no scientific literature on group decision making for designing complete streets. Moreover, policies and design guidelines usually deal with how to build complete streets rather than where to build them.

For many cities, localization decisions are made using a mono-criterion approach. This is also the case in Quebec City, where the engineering service usually chooses the streets that are to be rebuilt, based on an infrastructure obsolescence criterion. However, this engineering-centered approach has started to evolve, and Quebec City planners expressed interest in a new process to help better estimate a street's potential. In particular, the goal was to rank and identify higher priority streets that should be redesigned as complete streets in Quebec City [12]. As in most city administrations, transportation projects involve many departments such as transportation, engineering, urban planning and environment. This professional specialization entails language and knowledge boundaries that needed to be overcome in order to reconcile the different professional perspectives and reach a consensus [13].

The paper is organized as follows: Sect. 2 presents the methodology, Sect. 3 describes the decision process and the results, Sect. 4 contains a discussion, and Sect. 5 is the conclusion. Based on our experience with the case study, we describe the challenges encountered, the solutions applied, and explore other possible strategies to overcome similar hurdles in the future.

2 Methodology

Our proposed framework is a mix of the philosophy underlying Value Focused Thinking and MACBETH, a multi-criteria decision aid method that uses semantic pairwise comparisons to build interval scales, based on the difference of attractiveness between the alternatives performances on a given criterion [14]. We chose to use MACBETH since it is a user-friendly method supported by software. Another reason was the familiarity of the group members with aggregated scores based on a weighted mean and their wish to obtain a total preorder of the alternatives (streets). In addition, we could easily rank a large set of alternatives with MACBETH. Finally, as facilitators, we have an extensive experience with MACBETH and have used it in several research projects.

The framework consists of four phases (Fig. 1): (1) problem structuring; (2) MACBETH for the construction of criteria scales and of a common values system; (3) validation; and (4) ranking of the alternatives. Depending on the phase, different types of workshops and meetings were held. The SWS gathered professionals with a common background for shorter time periods to work on a specific issue related to their particular area of expertise. The inclusion of subgroup workshops (SWS) in addition to the usual GWS in MACBETH was helpful for preference modelling especially during criteria scales construction.

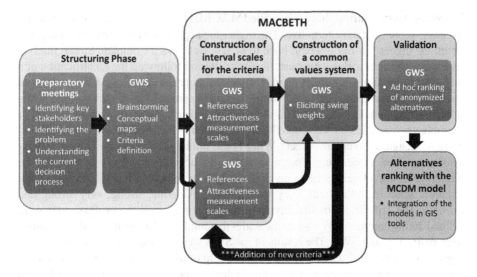

Fig. 1. Proposed framework

2.1 The Structuring Phase

The structuring phase involved two steps: decision problem structuring and MCDM evaluation model structuring [3]. The first step, completed in preparatory meetings, helped to understand the current decision environment, to define the problem and to scope the participation by identifying key stakeholders.

The second step served to structure the MCDM model by defining the criteria as proposed by Belton and Stewart [2]. During a first GWS, we led a brainstorm to identify the participants' main values, their requirements, needs, aspirations, issues and concerns in relation with the decision problem at hand. The questions posed during the brainstorming session were inspired by Keeney's VFT approach [15] and included questions such as: If there were only one street to rehabilitate, which one would you choose, and why? What would be an ideal street to be redesigned as a Complete Street? Which features make a street preferred to another street to be redesigned as a Complete Street? The answers were organized into dimensions reflecting observable attributes that could describe streets [16] and that would later serve as a basis for defining criteria. Similar dimensions were then grouped into categories. Subsequently, we used CmapTools to build a conceptual map linking the dimensions and the categories [17]. Finally, we presented and discussed the map results with the professionals. The emerging structure led to the construction of a final criteria set in the subsequent workshops. The usual alternative creation phase was not included in the process since the alternatives were naturally defined as the set of several thousand street segments in Quebec City, where a segment represents a portion of a street between two adjacent intersections with street types ranging from alleys to highways.

2.2 The MACBETH Phase

Following the structuring phase, we applied MACBETH to obtain an aggregated "attractiveness" score for each alternative. The higher the aggregated attractiveness score of a street, the higher priority it would have to be redesigned as a Complete Street. Interval level attractiveness scales were built for each criterion following the identification of a preference direction (minimize or maximize), appropriate echelons, and "good" and "neutral" reference points. A "good" reference level represents a satisfactory level while a "neutral" reference is a level that is considered neither attractive nor unattractive. For example, the participants were asked questions such as: what pedestrian flow is satisfying to design a Complete Street? What is the minimum acceptable human activity density required to design a Complete street? Participants then had to evaluate the difference in attractiveness between two echelons or reference levels using a 7-point semantic scale including null, very weak, weak, moderate, strong, very strong and extreme. The criteria scales that involved the expertise of all the professionals were defined during common GWS while the scales that implicated a specific field of expertise were defined during SWS. The results of the SWS were presented and explained at the beginning of every subsequent GWS with the help of the involved SWS professionals. The other professionals were then asked whether they agreed with the results and whether modifications should be made to the criteria scales and reference points. To standardize the scales, a common value system was then built using qualitative swing weights (scaling constants). These weights were obtained during GWS in a similar fashion to criteria scales, by comparing the difference of attractiveness between pair of fictitious alternatives using the previous semantic scales. There were as many fictitious alternatives as criteria. These alternatives were built in such a way that all the criteria had "neutral" scores except one criterion with the "good" reference as a score. For example, the professionals were asked to express their preference and the difference of attractiveness between a fictitious street A with a satisfying pedestrian flow ("good" score) and minimum acceptable characteristics for all other criteria ("neutral" scores), and a street B with a satisfying human activity density and minimum acceptable characteristics for all other criteria. After discussion and debate, a consensus was reached between the participants about the preference and differences of attractiveness. All the elicited information was captured in the M-MACBETH software to compute the attractiveness scales and the qualitative swing weights.

2.3 The Validation Phase

To validate the model, the participants were asked in the last GWS to rank a subset of anonymized alternatives. Their ad-hoc ranking was compared to the ranking results based on the aggregated alternatives' scores. When the alternatives were not in the same order as in the model, the participants had to explain their preferences. This new information was used to slightly adjust the model by modifying the difference of attractiveness between echelons of some criteria while ensuring consistency with the judgements provided by the stakeholders during the GWS and SWS.

2.4 The Alternatives Ranking Phase

Finally, the whole alternatives set was ranked using database management software and presented in a geographic information system (GIS). The integration of MCDM results in a GIS presented other challenges [18] that are not discussed in this paper.

3 Decision Process Description

The decision process for this project was organized according to the principles of decision conferencing [19]. It consisted of two preparatory meetings, six MCDM GWS and nine MCDM SWS conducted with 11 Quebec City professionals. The participants included an urban designer, a citizen's participation expert, a transportation engineer, an infrastructure engineer, a project manager, an environmental planner, three urban planners, a landscape architect and the sustainable development project director. The authors served as facilitators during the GWS and SWS. The workshops were carried out over two separate time periods. The first set of workshops took place between February and April 2016, and allowed us to structure the problem and to apply MACBETH using nine criteria. The second set of workshops took place during the month of July 2016 and led to the addition of two criteria, the modification of one criterion, and to model validation. The few months in-between were used for data processing and results production. Figure 2 shows the project's and workshops' timeline.

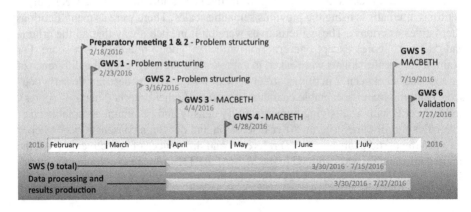

Fig. 2. Project and workshops timeline

The two preparatory meetings and the first two GWS served to structure the problem. From the third to the fifth GWS, MACBETH was applied to define the references and the criteria scales, and to elicit the criteria swing weights. At each of these workshops, new criteria were added in the model. Finally, during the sixth GWS, the model results were validated with the professionals. All the information was elicited through discussions and consensus between the professionals. Two to three facilitators were present during the GWS. The first facilitator animated the GWS, the second facilitator captured the information using the M-MACBETH software, and the second and the third facilitators advised the animator and analyzed the GWS. As for the SWS, they were held between the second and fifth GWS. Their main purpose was to define references and preference scales for specialized criteria that involved a small number of experts. The SWS were less formal than the GWS. During a SWS, the professionals had access to computers and could easily manipulate and explore several databases while the facilitator asked questions. In contrast, during a GWS, although a computer with a projector was available, it was difficult for the professionals to explore the data. The room layout did not allow for an easy access to the computer that was rather used to show the various phases and results live during the meetings. Furthermore, it did not have all the appropriate software to manage data. A whiteboard or a flipchart with pencils was always available to take notes and to write given information. We also made considerable use of Post-its and all the conversations were taped. The room configuration was in U-shape allowing the participants to have direct eye-to-eye contact and to easily see the different visual displays. The main differences between group and subgroup workshop features are presented in Table 1.

The GWS procedure was usually the same at each workshop. A workshop plan was presented to the professionals to explain what would be accomplished during the workshop. Afterwards, a brief summary of the previous group and subgroup workshop results was discussed. The professionals were given the opportunity to express themselves regarding specific points, if they felt it necessary. The methodology was then re-explained to the group and applied. At half-time, a break of approximately 15 min was normally enjoyed.

Table 1. Comparison of group workshop and subgroup workshop features

	Group workshop (GWS)	Subgroup workshop (SWS)
Length	Half a day (2 to 3 h)	Short (30 to 60 min)
Group size	8 to 11 professionals	1 to 4 professionals
Number of facilitators	2 to 3 facilitators	1 facilitator
Materials	Computer, projector, board or flipchart, pencil	Computer, board or flipchart, pencil
Expertise	Varied	Specific
Data access and manipulation	Hard	Easy
Main objectives	Different at each workshop	Define criteria references and preference scales

The GWS procedure was usually the same at each workshop. A workshop plan was presented to the professionals to explain what would be accomplished during the workshop. Afterwards, a brief summary of the previous group and subgroup workshop results was discussed. The professionals were given the opportunity to express themselves regarding specific points, if they felt it necessary. The methodology was then re-explained to the group and applied. At half-time, a break of approximately 15 min was normally enjoyed.

At the end of the process, 11 criteria along with their echelons, measurement units, good and neutral references, and attractiveness scales were constructed, their weights elicited, and ways to assess alternatives identified. Out of the 11 criteria, five involved quantitative scales, five used qualitative scales and one was composed of a qualitative and a quantitative scale. The number of echelons in the scales varied between 4 and 16. Seven criteria scales were defined during SWS. Normally, one scale was obtained during one SWS. However, the connectivity criterion required two SWS and the criteria of deprivation and urban tree canopy index were defined during the same SWS. Two additional SWS were held to explore data without defining any scales. Once all the criteria had been defined, the project director asked the other department directors whether they agreed with the criteria set. The department directors suggested the removal of some criteria and the addition of others. However, after deliberation, the participants decided to keep the criteria they chose and to add only one new criterion. They thought that the directors lacked the background since they were not present during the discussions where information and knowledge were exchanged.

The first MCDM model obtained prior to validation represented quite faithfully the group's preferences, as 80% of the alternatives ranked at ad hoc by the participants, were in the same order as that provided by the MCDM model. Subsequently, minor adjustment to the criteria scales were made to enhance the MCDM model. Table 2 presents the list of criteria, the criteria type, the number of echelons and the workshop type.

Table 2. Criteria description and type of workshop used

Criteria	Criteria type (number of echelons)	Workshop type
Safety	Qualitative (8)	Group
Connectivity	Quantitative (7)	Sub-group
Human activity density	Quantitative (6)	Sub-group
Deprivation	Qualitative (8)	Sub-group
Urban tree canopy index	Quantitative (5)	Sub-group
Bicycle network	Qualitative (5) and quantitative (6)	Sub-group
Street right of way	Quantitative (9)	Group
Pedestrian flow	Quantitative (7)	Sub-group
Citizen voice	Qualitative (4)	Sub-group
Urban planning	Qualitative (16)	Group
Bus network	Qualitative (10)	Group

4   Discussion

This project was a first experience in a MCDM workshop for all the stakeholders. They did not know what to expect beforehand. They had been invited by the sustainable development project director to participate, but few details about the methodology and the MCDM approach had been provided. They were surprised at the number of questions they had to answer during problem structuring and MCDM model building. It was quite difficult for them to think about the criteria and to express their preferences. Although identified as experts in their fields, they did not always have a deep knowledge of the data available and needed personal access to the data before they could express themselves. For several criteria, this was their first attempt at data interpretation. A lot of background work was needed to formalize their experience into useful information and ultimately to develop new and common knowledge. Furthermore, when the definition of a criterion scale engaged only one or two professionals with a specific expertise, the others felt excluded and had the impression that they were wasting their time since they could not meaningfully contribute to the discussion. A possible solution to overcome the data access problem in the future could be the inclusion of a data specialist with a dedicated computer during the GWS. A data specialist can explain the data quickly to the professionals during the workshop.

We decided to conduct SWS to overcome the difficulties encountered. During the smaller group meetings, the professionals had a better access to computers which made it easier for them to explore data. Moreover, it gave them the time required to think about data interpretation and to develop their thought process. After the first few SWS, the professionals had gained more confidence in the process. The less formal SWS settings may also have helped them to be more creative in their thinking. However, the use of SWS raises the question about the lack of discussion within the group as a whole. Even when a consensus is finally reached during GWS, the prior use of SWS could have lead to misunderstandings and consequently to a mistrust of the model results. This was of concern for us, since our view of an MCDM approach is that the process is often more important than the results. We were able to avoid this difficulty by ensuring at the beginning of each new GWS that the participants adhered to the results of the previous SWS.

The group size was another challenge. It created schedule and attendance problems. Even if the GWS were planned two to four weeks in advance so that everyone could attend, it was difficult to gather all the participants all the time. At each GWS, one or two professionals were missing. Some participants even joined the group later during the process as new participants or as substitutes for others. One reason is that the project continued during summer vacations. Many participants were then away. To overcome this problem, those who were present decided, by common agreement, to express their opinions to the best of their knowledge and to respect choices made in previous GWS.

As facilitators, it was challenging for us to manage the discussions. It is our opinion that in order to reach a consensus and to take into account all the perspectives, every participant should have a chance to express himself/herself. However, as is often the case, some individuals talked more and louder than others, and had undoubtedly a bigger influence on the end results. To minimize these aspects, we ensured that every professional expressed himself/herself and was in agreement before a final modelling decision was taken. Sometimes, we questioned the shyest participants directly to get their opinion. Once again, using SWS helped the process since the group sizes were smaller and allowed the participants to speak more and freely without peer pressure or fear of judgment. Off-subject discussions was another issue encountered as facilitators. Before going back to the main topic, we had to first ensure that there were no links between the topic and the off-subject.

The process's length of five months was another challenge. The duration is explained by several factors: The group size implied that more time was needed for discussions and negotiations to reach a consensus. Pairwise comparisons in MACBETH with 11 criteria involved many pairwise comparisons and sometimes lead to judgment inconsistencies that needed to be resolved. The size of the alternatives' set (many thousands) required that we build precise preference scales that accounted for all the possible performances. For example, the criterion of urban planning required the creation of a 16 echelon qualitative scale and took approximately 1h45 to define in a GWS. In fact, the definition of references and of echelons was usually the longest step in the process, and the construction of qualitative scales was usually longer than quantitative scales. This difference in the time required to construct scales can be explained by the fact that quantitative scales were usually readily available while qualitative scales required us to build them almost from scratch. The difficulty to gather all the professionals at the same time also contributed to lengthening the process. At one point, a snow storm forced the cancellation of a GWS that was reported several weeks later. Other issues related to the process's length were that facilitators had to re-explain the methodology at each GWS. Furthermore, three criteria required extensive data treatment that included computer programming.

The various difficulties encountered during the project raise the question of the appropriateness of a GWS structure. The workshops to define units and scales were perhaps too long, could be tiresome for the professionals while the results obtained at the end of a GWS were sometimes negligible. Despite half-day workshops, the professionals often felt exhausted at the end. To avoid this situation, long breaks were scheduled, and when after several discussions, it was still not possible to come to an agreement or to reach a consensus, small breaks of three to five minutes were taken, or another subject discussed. When a subject change occurred, the former subject was, depending on the case, discussed in a later GWS or in a SWS. With hindsight, we believe that morning workshops would probably have been less demanding for the participants than afternoon meetings.

5 Conclusion

The framework developed for this study allowed us to model group's preferences for sustainable transportation and complete streets in Quebec City by developing consensually shared views. The completed process resulted in the ranking of several thousand streets, as a function of their potential to become Complete Streets. Stakeholders from different backgrounds, multiple criteria, a consensus-building approach, and a large number of alternatives were all factors that amplified the complexity in preference modelling and increased the duration of the group decision process.

The challenges encountered because of the group's composition and size raise the question of a maximal or an ideal number of participating stakeholders in a GWS for consensus building. Eleven participants might have been too many for a method such as MACBETH, although still feasible. Nonetheless, the use of SWS was a good compromise to include more stakeholders and to build consensus without overburdening the process. It allowed us to overcome many of the difficulties encountered.

In the end, the application of the proposed framework proved to be a successful solution for bringing together individuals from different fields of expertise in a shared multi-criteria decision process. Our biggest challenge as facilitators was to manage the group's perception that we are the "experts" who had all the answers. We often had to remind them that this was their decision process, and that we were there to help them build their preference model and not to influence the process with our own biases. We were decision facilitators and not decision makers. Having said this, we, of course, intervened and made choices when the group was at a stalemate and needed to move forward. The participants seemed to be afraid of giving a wrong answer, and we had to stress the fact that there are no right or wrong answers in preference modeling. There are only answers reflecting the values of the individuals and the group. But, as the sustainable development project director stated: "The project brought the professionals out of their comfort zones."

Despite all the difficulties, the participants, and we, are very confident in the process's results and believe that this was a successful project. In March 2017, the Quebec City elected officials presented their Complete Street strategy to the media and the population. The cartographic tools developed with the model results were included as one of the key elements of their strategy. The results received an important media local coverage (newspapers and television). Furthermore, the mayor qualified the tool as the "quintessence of transdisciplinary" and the main municipal opposition party praised the project.

We are currently in the process of defining a follow-up project to extend our model and analysis from the 5,000 street segments that we ranked to all of the 20,000+street segments in Quebec City.

Acknowledgments. The authors wish to thank all the participants from Ville de Québec and UMR Sciences Urbaines for their input and valuable time. The authors are grateful for the financial support provided for this project by Thales, NSERC, MITACS, and Ville de Québec.

References

1. Belton, V., Stewart, T.J.: Multiple Criteria Decision Analysis: An Integrated Approach. Kluwer Academic Publishers, Boston (2002)
2. Belton, V., Stewart, T.: Problem structuring and multiple criteria decision analysis. In: Ehrgott, M., Figueira, J.R., Greco, S. (eds.) Trends in Multiple Criteria Decision Analysis, vol. 142, pp. 209–239. Springer, Heidelberg (2010)
3. Franco, L.A., Montibeller, G.: Problem structuring for multicriteria decision analysis interventions. Wiley Encycl. Oper. Res. Manag. Sci. (2010). doi: 10.1002/9780470400531.eorms0683
4. Belton, V., Pictet, J.: A framework for group decision using a MCDA model: sharing, aggregating or comparing individual information? J. Decis. Syst. **6**, 283–303 (1997). doi: 10.1080/12460125.1997.10511726
5. Alencar, L.H., de Almeida, A.T., Morais, D.C.: A multicriteria group decision model aggregating the preferences of decision-makers based on ELECTRE methods. Pesqui. Oper. **30**, 687–702 (2010). doi:10.1590/S0101-74382010000300010
6. Brans, J.-P., Mareschal, B.: Promethee methods. In: Multiple Criteria Decision Analysis: State of the Art Surveys. International Series in Operations Research & Management Science, vol. 78. Springer, New York (2005)
7. Escobar, M.T., Moreno-Jiménez, J.M.: Aggregation of individual preference structures in AHP-group decision making. Group Decis. Negot. **16**, 287–301 (2007). doi:10.1007/s10726-006-9050-x
8. Macharis, C., Turcksin, L., Lebeau, K.: Multi actor multi criteria analysis (MAMCA) as a tool to support sustainable decisions: state of use. Decis. Supp. Syst. **54**, 610–620 (2012). doi: 10.1016/j.dss.2012.08.008
9. Damart, S.: A cognitive mapping approach to organizing the participation of multiple actors in a problem structuring process. Group Decis. Negot. **19**, 505–526 (2010). doi:10.1007/s10726-008-9141-y
10. Haezendonck, E.: Introduction: transport project evaluation in a complex European and institutional environment. In: Transport Project Evaluation Extending the Social Cost-Benefit Approach. Edward Elgar, Cheltenham, Glos, UK; Northampton, MA, pp. 1–8 (2007)
11. McCann, B.: Completing Our Streets: The Transition to Safe and Inclusive Transportation Networks. Island Press, Washington (2013)
12. Marleau Donais, F., Lavoie, R., Abi-Zeid, I., Delisle, J.-P.: Évaluation du potentiel des rues à être aménagées en rues conviviales - Une approche par analyse multicritère. 61 (2016)
13. Franco, L.A.: Rethinking Soft OR interventions: models as boundary objects. Eur. J. Oper. Res. **231**, 720–733 (2013). doi:10.1016/j.ejor.2013.06.033
14. Bana e Costa, C.A., De Corte, J.-M., Vansnick, J.-C.: MACBETH. Int. J. Inf. Technol. Decis. Mak. **11**, 359–387 (2012). doi:10.1142/S0219622012400068

15. Keeney, R.L.: Developing objectives and attributes. In: Advances in Decision Analysis: From Foundations to Applications. Cambridge University Press, pp. 104–128 (2007)
16. Tsoukiàs, A.: From decision theory to decision aiding methodology. Eur. J. Oper. Res. **187**, 138–161 (2008). doi:10.1016/j.ejor.2007.02.039
17. Cmap. CmapTools. In: Cmap (2017). http://cmap.ihmc.us/cmaptools/. Accessed 9 Jan 2017
18. Ferretti, V., Montibeller, G.: Key challenges and meta-choices in designing and applying multi-criteria spatial decision support systems. Decis. Supp. Syst. **84**, 41–52 (2016). doi: 10.1016/j.dss.2016.01.005
19. Phillips, L.D.: Decision conferencing. In: Edwards, W., Miles, R.F.J., von Winterfeldt, D. (eds.) Advances in Decision Analysis: From Foundations to Applications, pp. 375–399. Cambridge University Press, Cambridge (2007)

A Group Decision Outranking Approach for the Agricultural Technology Packages Selection Problem

Pavel A. Álvarez Carrillo[1(✉)], Juan C. Leyva López[1],
and Omar Ahumada Valenzuela[2]

[1] Department of Economic and Management Sciences,
University of Occident, Culiacan, Mexico
{pavel.alvarez,juan.leyva}@udo.mx
[2] CONACYT Research Fellow, Management Sciences Doctorate Program,
University of Occident, Culiacan, Mexico
omar.ahumada@udo.mx

Abstract. The selection of a technological packages for agriculture is a complex task. Since selecting the best suited for a particular farm, given the rapid development of technology and the many combinations available, is a difficult problem for the decision maker (DM). This paper deals the selection problem of technological packages identifying criteria and alternatives in a group decision making process. The proposed model presents a multicriteria group decision model for ranking technology packages based on the outranking methods. This model is appropriate for those cases where there is great divergence among the DMs. The methodology can be used for defining rural credits (for adapted technological packages) in order to improve farmer's competitiveness and profitability.

Keywords: Group decision making · ELECTRE methods · Technology packages · Selection problem

1 Introduction

Managing a farm involves decision making in planning, implementation and monitoring, while contemplating the available resources and markets [1]. A big part of those decision, is crop selection and crop management (planting, cultivation, weed control, etc.), which is very important for reducing costs, improving productivity and achieving a profit in the very competitive and uncertain market for agricultural products.

Adopting new methods is usually hard for growers, since it comes at a risk, from the unknown results of new methods and the potential investment requirements, even when the potential benefits are significant [2]. The determinants for the adoption of new methods include factors such as an extension contact, education, farm size, credit availability, fertilizer use, low land area, yield and profitability [3].

By improving crop decision modelling, and technology selection (also known as technology package), which is often overlooked in farm management, growers can better tackle these complex issues. A technology package, usually consists in crop

© Springer International Publishing AG 2017
M. Schoop and D.M. Kilgour (Eds.): GDN 2017, LNBIP 293, pp. 187–201, 2017.
DOI: 10.1007/978-3-319-63546-0_14

variety, fertilizer, planting method and pest control, which are offered in an integrated way [4]. Selecting the best suited for a particular farm, given the rapid development of technology and the many combinations of technologies available, is a difficult task for the decision maker (DM). Particularly when considering that there could be many conflicting objectives for the different stakeholders (considered as DM) within and outside of the farm, for example, the grower might want to maximize profits, or satisfy the demand of a customer, while the bank that lends the farm money, wants to have the least risk possible in the farm business model, and the government demands from the farm to comply with environmental rules, such agrochemicals and fertilizer use. All of these seemingly different objectives might be affected by the selection of the above mentioned technology package.

These different perspectives might call for group decision making techniques, which can be defined as the reduction of different individual preferences in a given set, for a single collective preference [5]. The problem is to reach an agreement among the DMs, which may diverge in their perception of the problem and have different interests (as the example mentioned above), but all are responsible for the well-being of the organization and share responsibility for the decision implemented [6]. But this process is not conflict-free, and there may be some differences among the DMs, caused by a large number of factors, such as for example, different ethical or ideological beliefs, different specific objectives or different roles within the organization [7, 8].

Under the proposed method, DMs may exchange opinions and relevant information, but a group consensus is only needed to define a potential set of alternatives [9]. Then, each member defines his/her own criterion, the appropriate evaluations, parameters (weights, thresholds, etc.) and a multicriteria method is used to obtain their personal ranking. Thereafter, each decision-maker's criterion of the decision-maker is considered separately, and the information contained in his/her individual preference is aggregated into a final collective order, using this the same multicriteria decision approach [10].

The proposed model presents a multicriteria group decision model for ranking technology packages based on the outranking methods. This model is appropriate for those cases where there is divergence among individual results of DMs. When divergence is presented, the multicriteria group decision model supports DMs for redefinition of parameters regarding individual preference and collective preference. The model proposes new result remaining preference and matching better with collective solution. There are some consensus models supporting group decision making process to reach better consensus [11–13]. However, group aiding approaches based on Multicriteria Decision Analysis (MCDA) support feedback mechanism in preference representation [12, 14], or other different for disaggregate parameters as inter-criteria parameter. In [15] they asserted, definition of parameter for multicriteria methods is a complex task and it will more complex for group context. Parameter as preference representations is holistic parameter from DM, a feedback mechanism in more disaggregate level can be more complex for DM. In this sense, the advantage of the present approach supports those group decision making process where DM must define disaggregate preference as inter-criteria parameter.

Using the methodology commented above, we explore the technology package selection problem for a group of stakeholders (DMs) that need to reach a consensus solution, modelling it as for an instance of the group ranking problem using a multi-criteria group decision aiding methodology based on the outranking approach. In this context, DMs with different interests generated their own ranking of technology packages. Naturally, disagreements were found between the individual rankings and the collective solution. The approach uses a model for inferring inter-criteria parameters, where the DMs generate individual rankings that exhibit less disagreement with the collective solution [16].

2   Evaluation of Technological Packages

Although farmers have significantly increased the amount of food available for human consumption in the past century, the rate of productivity increases has slowed down in recent years [17]. This new trend is troubling, given that growers need to double crop production over the next 35 years, just to ensure that there is enough safe and nutritious food to feed a rapidly crowding planet [18]. Particularly important, for the sustainability of the activity, is that such increments, come from the development of new technology packages, not only from the utilization of more arable land in detriment of other uses.

Agricultural technologies are often presented as a package of interrelated technologies for example, high yielding seeds, fertilizer, herbicides, and chemicals. [19]. Accordingly, one major focus in the literature in recent years has been the investigation of the decision-making process characterizing choice of the optimal combinations of the components of a technological package over time [20], but empirical evidence suggest that for the most part, farmers choose to adopt inputs sequentially, adopting initially only a part of the package and subsequently adding components over time [21]. Incremental adoption of technology packages is reasonable, in terms of the uncertain outcome of the new technology [22]. It seems there is evidence that farmers prefer to have sequential adoption of technologies, even when there is a high interaction between the individual components, due to the risk in terms of investment and the learning curve required by some of the technologies [4].

However, better results could be obtained by adopting an integrated package, for example a new seed, may be further improved by applying appropriate levels of fertilizer and irrigation or modifying plant density [21]. Adoption studies of technologies for farmers have considered single innovations in isolation ignoring the process of adoption among a set of components of technological package [23], but new developments such as genetically modified seed varieties often work in tandem with other innovations, such as herbicides that improve yields considerably [24].

An appropriate model to select the most suitable technological package given their particular benefits, both in terms of yield, but more importantly in their return on investment (given the often high cost of new technological improvements) is an important problem that needs to be tackled [3], the present research, aims to improve technology selection decisions and the adoption of those technologies in order to improve farmers competitiveness and profitability [20].

3 Group Decision Making Approach for ELECTRE Method

The group decision-making (GDM) is an interactive important process that can be supported by a variety of GDM methods [25] and GDM consensus process. For the evaluation of technological packets, we use methods based on outranking approach. Particularly, we use ELECTRE III [7, 26] to generate individual preferential model represented by a valued outranking relation (S_A^σ) and an outranking method for group decision-making based on ELECTRE to generate a collective preferential model represented by a collective outranking relation ($S_A^{\sigma,G}$). Due the strong influence ELECTRE exerts on the last method, it is called ELECTRE-GD [8]. After construction of the preferential model stage, the exploitation of model is the subsequent stage. Those two stages are described in the following sections.

The outranking approach is recognized in literature to be useful for real problems. Outranking methods show ability to deal with ordinal and more or less descriptive information on the alternative plans to be evaluated [27]. Outranking approach deal with decision making problems without any normativity, because its constructive approach more flexibility is open to DMs express their preference.

3.1 Constructing the Individual Preferential Model with ELECTRE III

In multicriteria decision aids, models concern preference of decision-maker as a very important input data in order to construct the preferential model. In this section we briefly describe the ELECTRE III method. ELECTRE performs pairwise comparison between alternatives (actions) assessing the asseveration "action a is at least as good as action b", if this asseveration is true, we can say "a outranks b" and is denoted by aSb.

Comprehensive concordance index Let $A = \{a_1, a_2, \ldots, a_m\}$ be the set of decision alternatives or potential actions, and it is required to define a coherent family of criteria $g_j, j = 1, 2, \ldots, n$. For each criterion the concordance and discordance indices are calculated. The comprehensive concordance index $C(a_i, a_l)$ measures the performances on all criteria of the pair of alternatives $(a_i, a_l) \in A \times A$ to evaluate in what grade the criteria support the assertion "a_i outranks a_l".

$$C(a_i, a_l) = \frac{\sum_{j=1}^{n} w_j \cdot C_j(a_i, a_l)}{\sum_{j=1}^{n} w_j} \tag{1}$$

where w_j is the relative weight of the criteria g_j. $C_j(a_i, a_l)$ is the partial concordance index to evaluate the grade criterion g_j supports the assertion "a_i outranks a_l". The discordance index $d_j(a_i, a_l)$ measure how much is discordant the criteria g_j with the assertion "a_i outranks a_l".

The fuzzy outranking relation The model building phase combines those two previous measures to produce a measure of the degree of outranking, that is, a credibility index $\sigma(a_i, a_l)$; $(0 \le \sigma(a_i, a_l) \le 1)$ which assesses the strength of the assertion that "a_i is at least as good as a_i", $a_i S a_l$. The credibility degree for each pair $(a_i, a_l) \in A \times A$ is defined as follows:

$$\sigma(a_i, a_l) = \begin{cases} C(a_i, a_l) & \text{if } \overline{F}(a_i, a_l) = 0 \\ C(a_i, a_l) \times \prod_{j \in \overline{F}(a_i, a_l)} \frac{1 - d_j(a_i, a_l)}{1 - C(a_i, a_l)} & \text{if } \overline{F}(a_i, a_l) \neq 0 \end{cases} \quad (2)$$

where $\overline{F}(a_i, a_l) = \{j \in F / d_j(a_i, a_l) > C(a_i, a_l)\}$, the set of criteria where discordance index is higher than comprehensive concordance index.

3.2   Constructing the Collective Preferential Model with ELECTRE GD

The ELECTRE GD method is based on a natural heuristic based on majority rules combined with respect to significant minorities. For the GDM problem the group is composed for $M = \{1, 2, \ldots, r\}$.

Let $\sigma_k : A \times A \rightarrow [0, 1]$ be a valued binary relation, which aggregates the preferences of the kth member on the multiple criteria describing the elements of A. Let O_k be a complete ranking of A derived using some procedure for exploiting σ_k.

At the first stage, the method identifies conflicts between preferential model of DM$_k$ (σ_k) and its corresponding ranking O_k, constructing what it is called preference matrix. The basic idea is considering each member k as a criterion of the multicriteria problem and each pair of action $(a_i, a_l) \in A \times A$ should be compared according to the point of view of criterion k. The preference matrix asserts the asseveration a_i is at least as good as a_j.   The   preferential   model   agrees   with   this   asseveration   when $\sigma_k(a_i, a_l) \ge \lambda$; $(0 \le \lambda \le 1)$, and disagree when $\sigma_k(a_i, a_l) \le \lambda - \beta$ (β is a threshold parameter). The ranking O_k express how preferred is a_i over a_l.

The ELECTRE-GD method works based on concordance and discordance principles. Action a_i outranks action a_l from the point of view of actor k, it is defined as a restricted outranking relation $a_i S a_l$. An actor k is in concordance with the assertion $a_i S_G a_l$ (S_G means group outranking), if and only if $a_i S a_l$. Thus $C(a_i, a_l)$ denote the *concordant coalition*, the set of actors, which are in concordance with $a_i S_k a_l$. An actor k is in discordance with the assertion $a_i S_G a_l$ if and only if a_l is strictly preferred to a by actor k. $d(a_i, a_l)$ denote the *discordant coalition* when actors are discordant with $a_i S_G a_l$. In case the number of actors discordant with $a_i S_G a_l$ increase, a veto coalition is determined to veto $a_i S_G a_l$ in case presence of strong discordance.

The comparability index $r(a_i, a_l)$ considers the credibility degree of the group outranking $a_i S_G a_l$. A decision is valid only if an important part of the group votes effectively (50% is a usual threshold). The fuzzy outranking relation of ELECTRE-GD is calculated based on previous index as follows.

$$\sigma_G : A \times A \rightarrow [0, 1],$$
$$\sigma_G(a_i, a_l) = C(a_i, a_l) \cdot (1 - d(a_i, a_l)) \cdot r(a_i, a_l) \tag{3}$$

3.3 Exploitation Phase of the Preferential Model

In multicriteria decision aids, outranking methods construct a preferential model usually expressed as a fuzzy outranking relation. Very important methods are highlighted in literature, the net flow function [28, 29] and distillation methods [26]. A main problem with those methods, they lack of heuristics to minimize inconsistency between preferential model and generated ranking. If the preferential model of DM_k express $\sigma_k(a_i, a_l) \geq \lambda$ it means $a_i S_k a_l$, but if a_i is worse ranked than a_l, it is considered not "well-ordered". It is desirable to use methods where inconsistency is minimized. For the package technology evaluation, the exploitation method is an multiobjective evolutionary algorithm [30], which optimizes three main objective functions in order to reduce inconsistency and incomparability and increase the credibility level of a crisp outranking relation $S_A^\lambda, (0 \leq \lambda \leq 1)$.

3.4 Group Decision Making Process Based in Individual Outranking Results

The GDM process implemented for evaluate technological packages is based on individual results generated by DMs. Those results are a fuzzy outranking relation and an individual ranking. There are two main methods which generate those output formats ELECTRE-III [26] and PROMETHEE [31]. The GDM process implemented in this work follows the main idea of derivate a collective preference model from those individual preferential model and individual rankings. In this sense, ELECTRE-GD accomplish with the task of proposing a collective solution based on the individual results.

In the schema of Fig. 1 seven steps are represented in preference expressions by DMs. For this GDM process a main requirement is DMs agree with a set of potential alternatives. After that, in step 1 every DM define his/her own set of criteria and a performance matrix is filled based on the agree alternatives and his/her own criteria (step 2). The step 3 allow DMs express their individual preferences with inter-criteria parameters weight (w), indifference threshold (q), preference threshold (p) and veto threshold (v). In the step 4, DMs use ELECTRE III to generate their preference model in a valued outranking relation (σ_k). For the step 5 the exploitation stage is performed by the evolutionary algorithm developed by [30]. The step 6 use the ELECTRE-GD to integrate individual valued outranking relation and construct the collective preferential model. Step 7 allows generate the collective ranking by the exploitation of the collective preferential model with Leyva, Aguilera [30].

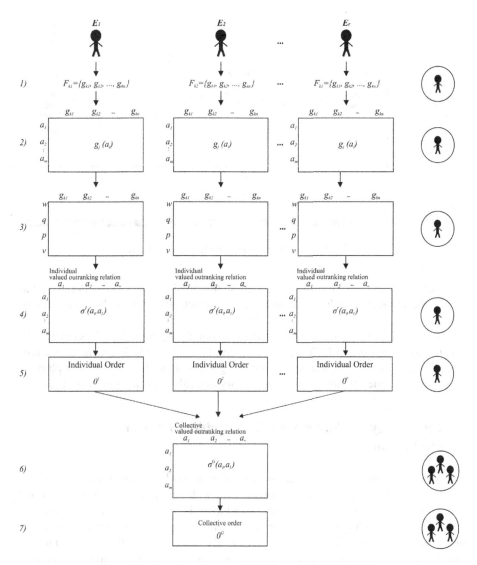

Fig. 1. Schematic process of GDM using ELECTRE-GD

3.5 The Web Based System for Group Decision Making Process

SADGAGE [32] is a web-based decision support system for multicriteria ranking problems. SADGAGE supports the group decision making process and it follows some important elements to support the structuration of the problem (See Fig. 2), adequate integration of preference and a guide to reach certain consensus level. Figure 2 shows the agenda where the multicriteria ranking problem is developed though a set of more specific issues (Topics), each one requiring a decision-making process. Each process consists of one or more zones. Four different zones come in the following temporal

Set schedule

[Show my meetings list | Configurate | Set tools | Lead the meeting | See the preview]

Edit schedule

Project: GDN 2017

Edit	No.	Topic	Start date	Start time	End date	End time	Status	Coordination
Modificar	1	First touch with the facilitator	22/01/2017	02:30	22/01/2017	02:30	Not started	Sequential
Modificar	2	Problem description	22/01/2017	02:30	22/01/2017	02:30	Not started	Sequential
Modificar	3	Choose a coherent family of criteria	23/01/2017	17:40	23/01/2017	17:40	Finished	Parallel
Modificar	4	Generate alternatives	23/01/2017	17:40	23/01/2017	17:40	Finished	Sequential
Modificar	5	Choose an stable family of alternatives	23/01/2017	17:40	23/01/2017	17:40	Finished	Sequential
Modificar	6	Assess alternatives for each criterion	24/01/2017	01:30	24/01/2017	01:30	Finished	Parallel
Modificar	7	Assign weights for each criterion	24/01/2017	11:20	24/01/2017	11:20	Finished	Parallel
Modificar	8	Establish indifference and preference thresholds for each criterion	24/01/2017	16:00	24/01/2017	16:00	Finished	Parallel
Modificar	9	Modeling individual decision maker's preferences - ELECTRE-III	24/01/2017	16:10	24/01/2017	16:10	In process	Parallel
Modificar	10	Modeling group's preferences - ELECTRE-GD	24/01/2017	17:00	24/01/2017	17:00	Finished	Parallel
Modificar	11	Voting for the final solution					Not started	Parallel

1

Fig. 2. Agenda for the multicriteria ranking problem

order: divergent (search for information); groan (discuss issues); convergent (attempt to reduce the number of solutions); and closure (select one solution by consensus or voting).

The decision process described in follow section is supported by SADGAGE [32], which implements the ELECTRE-III and ELECTRE-GD methods for the GDM process. A variant of the implemented outranking methods is that exploitation phase is developed with the multiobjective evolutionary algorithm pointed out in Sect. 3.3.

4 Technology Packages Selection Problem for Corn Production

The corn producers in Sinaloa State of Mexico are funded by a trust fund named FIRA (from the Spanish "instituted trust funds related with agriculture"). FIRA defines just one "standard" technological package (TP) which it is mainly based on the traditional methods, and includes the cost of activities related with land preparation, planting, management and harvesting. Even when the lender does not force the grower to follow exactly the TP generated by FIRA, it is the basis for estimating the financial requirements of growers during the agricultural year, and for granting loans (fixed quantity per hectare) to corn producers. The problem is that FIRA construct just one TP based on experience of past plantings, but there are many different emerging technologies that farmers could implement. For the present work, the aim is to generate a set of technological packages, that could be evaluated by a group of experts to select the TP that better matches the objectives of group of experts. Hereafter, we use the terms expert and decision-maker (DM) indistinctly.

The group of DMs is constituted by 4 members representing different sector or organizations. DM1 represents FIRA organization, DM2 and DM3 represent corn producers and DM4 serves as an environmentalist because concerns damage environment.

They expressed their points of view in meetings and listed the factors they perceived important to consider for certain practices in corn crop production. The decision process was supported by the multicriteria group decision support system SADGAGE [32], which implements the ELECTRE-III and ELECTRE-GD methods for the GDM process. The GDM process at SADGAGE follows some important elements support the structuration of the problem, adequate integration of preference and a guide to reach certain consensus level.

The main goal of the group of DMs is not to reach a high level of consensus, but to obtain a set of TPs, evaluate them and then identify the member's preferences to reach a consensus on the preferred technology packages to generate new sustainable credits more related with corn producer needs and taking care of the environment in the region.

The DMs hold a meeting with the facilitator-analyst in order to identify the criteria to be used in the selection process. We now show, the criteria definition for technology package for corn production. The experts proposed six criteria described in the following analytical way. Criteria related with cost are measured with the national currency (Mexican pesos).

Crop yields (C1): It is the performance of the seed once the crop is harvested. It can be influenced by climatological states, but the management of the crop is important to reach it estimated performance. The crop yield is measured in quantity of tons per hectare harvested.

Soil preparation (C2): It is the process where the land is prepared to the next stage, planting. It includes soil tests, tracking, seed brand, canals, scarification, canals cleaning, irrigation for preparation. The soil preparation is a set of activities which impact on the cost, in this sense the application of those activities is measured amount of money required.

Planting (C3): The planting stage of the crop includes; the seed, sow the seed, seed insecticide treatment, starter fertilizer, estimation of other activities as prevention for risks of the seed. The activities related with planting impact on the cost of the crop, in this sense the application of those activities is measured amount of money required.

Fertilization (C4): It is the application of some product to soil or to plant tissues (usually leaves) to supply one or more plant nutrients essential to the growth of plants. Some substances are Urea and ammonia and the way they are applied to the soil or plant. It is measured amount of money required.

Pest control (C5): It is the set of activities that controls weeds, plagues. Some products are applied by air or ground. It is measured amount of money required.

Harvest time (C6): It is the time of maturation of the plant to be in condition of harvesting. As corn presents a diver variety of plants, each region of the country uses different kind of maturation. In Sinaloa state, there are three common types of see to be sown in the region related with its maturity. Early intermediate since 180 to 185 days, Intermediate since 190 to 195 and Late intermediate since 195 to 200. The evaluation of this criterion is related with the level of maturation where 1 is early intermediate, 2 is intermediate, and 3 is late intermediate.

At Table 1 the performance matrix presents six TPs and the above defined criteria. As consequence of the iteration of experts with analyst, the elicitation of the inter-criteria parameters results are shown in Table 2.

Table 1. Performance of technological packages

	C1: Crop yield	C2: Soil preparation	C3: Planting	C4: Fertilization	C5: Pest control	C6: Harvest time
TP A	10	3534	7000	6321	1150	1
TP B	11	3534	6725	6003	900	1
TP C	11.7	3234	6275	5733	883	2
TP D	11.7	2834	6725	5700	800	2
TP E	12.3	3234	6900	5664	850	3
TP F	13	2834	6275	5328	883	3

Table 2. Inter-criteria parameters: weights and thresholds

		C1	C2	C3	C4	C5	C6
DM1	Dir	Max	Min	Max	Max	Max	Min
	w	26.1	11.95	11.95	11.95	11.95	26.1
	q	0	0	0	0	0	0
	p	0.6	300	175	69	50	1
	v						
DM2	Dir	Max	Min	Max	Max	Max	Max
	w	28.55	4.75	19.05	14.25	9.55	23.85
	q	0.1	300	100	200	100	0
	p	0.6	400	275	339	267	1
	v	1.3					
DM3	Dir	Max	Min	Min	Max	Min	Min
	w	30	10	10	16.7	10	23.3
	q	0.1	0	100	69	100	0
	p	0.6	300	175	270	267	1
	v	1.7					
DM4	Dir	Max	Min	Max	Min	Min	Min
	w	2.45	20.75	14.65	26.8	26.8	8.55
	q	0.6	0	100	36	0	0
	p	1	300	450	200	100	1
	v				500	250	

Having finished step 3 of the schema of the Fig. 1. At step 4, the ELECTRE III method is used to generate the preferential model in the format of a fuzzy outranking relation (σ_k) for every DM_K. Once we obtained the individual preferential model a MOEA is used for the exploitation phase, and generate an individual ranking.

10 ranking were generated for each DM to propose robust individual solutions. Table 3 (a) shows a summary of weighed (WS) sum of alternatives appearing in each of the six position of the ranking for every DM. Every time an alternative appears in the position one, a score of value 6 is accumulated, the weighed sum method is applied. Last position of the ranking assigns a score of value 1. The WS is used to propose an individual ranking based on the previous robust analysis. The information of Table 3(a) suggests the following ranking for each DM on Table 3(b).

Table 3. Summary of weighed sum alternatives in every position of ranking for each DM

<table>
<tr><td colspan="7">a) Summary of weighed</td></tr>
<tr><td></td><td>TPA</td><td>TPB</td><td>TPC</td><td>TPD</td><td>TPE</td><td>TPF</td></tr>
<tr><td>DM1</td><td>59</td><td>42</td><td>34</td><td>32</td><td>12</td><td>31</td></tr>
<tr><td>DM2</td><td>31</td><td>10</td><td>24</td><td>36</td><td>57</td><td>52</td></tr>
<tr><td>DM3</td><td>15</td><td>42</td><td>44</td><td>38</td><td>15</td><td>56</td></tr>
<tr><td>DM4</td><td>10</td><td>20</td><td>30</td><td>57</td><td>40</td><td>53</td></tr>
</table>

b) Rankings for each DM

DM1: TPA ≻ TPB ≻ TPC ≻ TPD ≻ TPF ≻ TPE
DM2: TPE ≻ TPF ≻ TPD ≻ TPA ≻ TPC ≻ TPB
DM3: TPA ≻ TPB ≻ TPC ≻ TPD ≻ TPE ≻ TPF
DM4: TPD ≻ TPF ≻ TPE ≻ TPC ≻ TPB ≻ TPA

Once we obtained the individual ranking (step 5) we can continue to the step 6 of the GDM process. The ELCTRE-GD use as input the preferential model (σ_k) and ranking (R_k) of every DM. In the step 6 of the GDM process, ELECTRE-GD constructs a collective fuzzy outranking relation (σ_G). Once we obtained his valued matrix the exploitation phase is again performed and the collective ranking is generated (step 7).

Table 4 shows 10 collective rankings generated from the collective fuzzy outranking relation (σ_G). We can find some smooth inversions between rankings. The weighted sum helps us to propose a collective ranking based on the robust analysis of Table 4. The analysis suggests the following collective ranking:

TPF ≻ TPD ≻ TPC ≻ TPB ≻ TPE ≻ TPA

Table 4. Different collective rankings for group DMs

<table>
<tr><td colspan="11">G</td></tr>
<tr><td></td><td>R^1</td><td>R^2</td><td>R^3</td><td>R^4</td><td>R^5</td><td>R^6</td><td>R^7</td><td>R^8</td><td>R^9</td><td>R^{10}</td></tr>
<tr><td>1</td><td>TPF</td><td>TPF</td><td>TPF</td><td>TPF</td><td>TPF</td><td>TPF</td><td>TPF</td><td>TPD</td><td>TPF</td><td>TPD</td></tr>
<tr><td>2</td><td>TPD</td><td>TPD</td><td>TPD</td><td>TPD</td><td>TPC</td><td>TPD</td><td>TPC</td><td>TPF</td><td>TPD</td><td>TPF</td></tr>
<tr><td>3</td><td>TPC</td><td>TPC</td><td>TPC</td><td>TPC</td><td>TPD</td><td>TPC</td><td>TPD</td><td>TPC</td><td>TPC</td><td>TPC</td></tr>
<tr><td>4</td><td>TPB</td><td>TPB</td><td>TPB</td><td>TPE</td><td>TPB</td><td>TPB</td><td>TPB</td><td>TPE</td><td>TPE</td><td>TPA</td></tr>
<tr><td>5</td><td>TPE</td><td>TPE</td><td>TPE</td><td>TPB</td><td>TPE</td><td>TPA</td><td>TPA</td><td>TPA</td><td>TPB</td><td>TPE</td></tr>
<tr><td>6</td><td>TPA</td><td>TPA</td><td>TPA</td><td>TPA</td><td>TPA</td><td>TPE</td><td>TPE</td><td>TPB</td><td>TPA</td><td>TPB</td></tr>
<tr><td>λ</td><td>0.612</td><td>0.612</td><td>0.612</td><td>0.631</td><td>0.617</td><td>0.596</td><td>0.6</td><td>0.642</td><td>0.631</td><td>0.618</td></tr>
</table>

A sensitive analysis could be performance making small variation on individual ranking for one and/or two DMs at the same time. If the solution is robust, it is expected to present minimal changes in the collective solution. The proposal of solution was generated from divers generated collective solutions. In this particular case, just a robust analysis was considered.

A summary of final individual ranking and collective ranking are showed in Table 5. For every individual ranking the number of rank reversals was identified. We can see those differences between individual and collective ranking as disagreements. A proximity index named weighted Kendall version was calculated based on those rank reversals [33]. Based on the proximity index we computed a consensus level (0.612). Even when the consensus level can be considered low, the preferences of DM and their differences in opinion are shown in the comparison between their individual ranking against the collective ranking.

Table 5. Individual and collective ranking

Position	DM1	DM2	DM3	DM4	Collective
1	TPA	TPE	TPF	TPD	TPF
2	TPB	TPF	TPC	TPF	PTD
3	TPC	TPD	TPB	TPE	TPC
4	TPD	TPA	TPD	TPC	TPB
5	TPF	TPC	TPA	TPB	TPE
6	TPE	TPB	TPE	TPA	TPA
Disagrees	11	6	3	3	
Proximity	0.383	0.533	0.75	0.783	
Consensus level (C_A) 0. 612					

The selected TP is the best balance between the preferences of the DMs, which makes it easier to implement once, the group has settled for one alternative. This greatly reduces "buyer's remorse" from the DMs, easing the conflicts in the implementation and during the evaluation of the results for the year. Particularly for those DMs whose championed alternative, was not picked, but it turns out better than the implemented solution.

That the selected alternative is turns out not to be the best solution after the season, is one of the limitations of the methodology, since it does not guarantee that the consensus decision is the best for satisfying all the criteria, particularly for those that involve uncertainty, or depend on random variables, which is the case for growers, who depend on the results from uncertain yields, prices and weather, thus there is no certainty that the selected alternative is the best for the coming season.

5   Conclusions

The selection problem treated in this work helped to define six technological packages and perform a multicriteria evaluation. A group decision making process was performed, implementing ELECTRE methods for individual stages and the integration of individual preference stage (group context). The group decision making process supported the DMs to generate robust individual ranking and collective ranking. Some important disagreements were found between two DMs against collective ranking (DM1 and DM2). The analysis helped them to understand the main differences on their points of view, with regards to corn technology package selection. We found that the most disagreement with the collective ranking came from the lending official (DM1), which has a priority in reducing risk, so that the loan may be repaid, without considering potential profit. This point of view does not always match the point of view of the corn producers, since they are usually not as risk averse as lending officers, and some of them might have higher risk profiles.

As mentioned before, one of the main benefits of the method is the support for implementation, given the transparency of the proposed method, which helps the group get behind their consensus alternative, even if it turns out it was not the best solution given the characteristics of that year.

Another way the consensus helps the implementation process, is that it gives those putting it together, a mandate from the DMs, and they can be confident that the selected course of action has the support of key DMs, and improves the willingness that key areas are working together.

These findings show the necessity of the trust fund to collaborate closely with corn producers to define appropriate credits.

References

1. Kahan, D.: Market-oriented farming: an overview. Food and Agriculture Organization of the United Nations (2013)
2. Leathers, H.D., Smale, M.: A Bayesian approach to explaining sequential adoption of components of a technological package. Am. J. Agr. Econ. 73(3), 734–742 (1991)
3. Kafle, B.: Determinants of adoption of improved maize varieties in developing countries: A review. Int. Res. J. Appl. Basic Sci. 1(1), 1–7 (2010)
4. Byerlee, D., de Polanco, E.H.: Farmers' stepwise adoption of technological packages: evidence from the Mexican Altiplano. Am. J. Agr. Econ. 68(3), 519 (1986). doi:10.2307/1241537
5. Jelassi, T., Kersten, G., Zionts, S.: An Introduction to Group Decision and Negotiation Support, pp. 537–568 (1990). doi:10.1007/978-3-642-75935-2_23
6. Khan, M.S., Quaddus, M.: group decision support using fuzzy cognitive maps for causal reasoning. Group Decis. Negot. 13(5), 463–480 (2004). doi:10.1023/B:GRUP.0000045748.89201.f3
7. Roy, B.: Multicriteria Methodology for Decision Aiding. Kluwer Academic Publishers, The Netherlands (1996)

8. Leyva, J.C., Fernández, E.: A new method for group decision support based on ELECTRE III methodology. Eur. J. Oper. Res. **148**(1), 14–27 (2003). doi:10.1016/s0377-2217(02)00273-4

9. Parreiras, R.O., Ekel, P.Y., Morais, D.C.: Fuzzy set based consensus schemes for multicriteria group decision making applied to strategic planning. Group Decis. Negot. **21**(2), 153–183 (2012). doi:10.1007/s10726-011-9231-0

10. Palomares, I., Estrella, F.J., Martínez, L., Herrera, F.: Consensus under a fuzzy context: Taxonomy, analysis framework AFRYCA and experimental case of study. Inf. Fusion **20**, 252–271 (2014). doi:10.1016/j.inffus.2014.03.002

11. Dong, Y., Li, C.-C., Xu, Y., Gu, X.: Consensus-based group decision making under multi-granular unbalanced 2-tuple linguistic preference relations. Group Decis. Negot. **24**(2), 217–242 (2015). doi:10.1007/s10726-014-9387-5

12. Dong, Y., Zhang, H.: Multiperson decision making with different preference representation structures: a direct consensus framework and its properties. Knowl.-Based Syst. **58**, 45–57 (2014). doi:10.1016/j.knosys.2013.09.021

13. Zhang, B., Dong, Y., Xu, Y.: Multiple attribute consensus rules with minimum adjustments to support consensus reaching. Knowl.-Based Syst. **67**, 35–48 (2014). doi:10.1016/j.knosys.2014.06.010

14. Herrera-Viedma, E., Herrera, F., Chiclana, F.: A consensus model for multiperson decision making with different preference structures. IEEE Trans. Syst. Man Cybern. Part A: Syst. Hum. **32**(3), 394–402 (2002). doi:10.1109/tsmca.2002.802821

15. Chakhar, S., Saad, I.: Incorporating stakeholders' knowledge in group decision-making. J Decis Syst **23**(1), 113–126 (2014). doi:10.1080/12460125.2014.865828

16. Álvarez, Pavel A., Morais, Danielle C., Leyva, Juan C., Almeida, Adiel T.: A multi-objective genetic algorithm for inferring inter-criteria parameters for water supply consensus. In: Gaspar-Cunha, A., Henggeler Antunes, C., Coello, C.C. (eds.) EMO 2015. LNCS, vol. 9019, pp. 218–233. Springer, Cham (2015). doi:10.1007/978-3-319-15892-1_15

17. Fan, M., Shen, J., Yuan, L., Jiang, R., Chen, X., Davies, W.J., Zhang, F.: Improving crop productivity and resource use efficiency to ensure food security and environmental quality in China. Journal of experimental botany, err248 (2011)

18. Grose, T.K.: The next GREEN revolution. ASEE Prism **25**(4), 28–31 (2015)

19. Aldana, U., Foltz, J.D., Barham, B.L., Useche, P.: Sequential adoption of package technologies: the dynamics of stacked trait corn adoption. Am. J. Agr. Econ. **93**(1), 130–143 (2011). doi:10.1093/ajae/aaq112

20. Feder, G., Umali, D.L.: The adoption of agricultural innovations: a review. Technol. Forecast. Soc. Chang. **43**(3), 215–239 (1993)

21. Leathers, H.D., Smale, M.: A Bayesian approach to explaining sequential adoption of components of a technological package. Am. J. Agr. Econ. **73**(3), 734 (1991)

22. Marra, M., Pannell, D.J., Ghadim, A.A.: The economics of risk, uncertainty and learning in the adoption of new agricultural technologies: where are we on the learning curve? Agric. Syst. **75**(2), 215–234 (2003)

23. Feder, G., Just, R.E., Zilberman, D.: Adoption of agricultural innovations in developing countries: a survey. Econ. Dev. Cult. Change **33**(2), 255–298 (1985)

24. Scandizzo, P.L., Savastano, S.: The adoption and diffusion of GM crops in United States: a real option approach. (2010)

25. Herrera, F., Martínez, L., Sánchez, P.J.: Managing non-homogeneous information in group decision making. Eur. J. Oper. Res. **166**(1), 115–132 (2005). doi:10.1016/j.ejor.2003.11.031

26. Roy, B.: The outranking approach and the foundations of ELECTRE methods. In: Bana e Costa, C.A. (ed.) Reading in Multiple Criteria Decision Aid, pp. 155–183. Springer, Berlin (1990)

27. Kangas, A., Kangas, J., Pykalainen, J.: Outranking methods as tools in strategic natural resources planning. Silva Fennica **35**(2), 215–227 (2001). doi:10.14214/sf.597
28. Behzadian, M., Kazemzadeh, R.B., Albadvi, A., Aghdasi, M.: PROMETHEE: a comprehensive literature review on methodologies and applications. Eur. J. Oper. Res. **200**(1), 198–215 (2010). doi:10.1016/j.ejor.2009.01.021
29. Macharis, C., Brans, J.P., Mareschal, B.: The GDSS PROMETHEE procedure - a PROMETHEE-GAIA based procedure for group decision support. J. Decis. Syst. **7**, 283–307 (1998)
30. Leyva-Lopez, J.C., Aguilera-Contreras, M.A.: A multiobjective evolutionary algorithm for deriving final ranking from a fuzzy outranking relation. In: Coello Coello, C.A., Hernández Aguirre, A., Zitzler, E. (eds.) EMO 2005. LNCS, vol. 3410, pp. 235–249. Springer, Heidelberg (2005). doi:10.1007/978-3-540-31880-4_17
31. Brans, J.P., Vincke, P.: A preference ranking organization method. Manage. Sci. **31**, 647–656 (1985)
32. Leyva López, J.C., Álvarez Carrillo, P.A., Gastélum Chavira, D.A., Solano Noriega, J.J.: A web-based group decision support system for multicriteria ranking problems. Oper. Res. Int. J., 1–36 (2016). doi:10.1007/s12351-016-0234-0
33. Leyva López, L., Alvarez Carrillo, P.A.: Accentuating the rank positions in an agreement index with reference to a consensus order. Int. T. Oper. Res. **22**(6), 969–995 (2015). doi:10.1111/itor.12146

Can the Holistic Preference Elicitation be Used to Determine an Accurate Negotiation Offer Scoring System? A Comparison of Direct Rating and UTASTAR Techniques

Ewa Roszkowska[1], Tomasz Wachowicz[2($\boxtimes$)], and Gregory Kersten[3]

[1] Faculty of Economy and Management, University of Bialystok,
Warszawska 63, 15-062 Bialystok, Poland
erosz@o2.pl
[2] Department of Operations Research, University of Economics in Katowice,
1 Maja 50, 40-287 Katowice, Poland
tomasz.wachowicz@ue.katowice.pl
[3] John Molson School of Business, Concordia University, Montreal, Canada
gregory.kersten@concordia.ca

Abstract. In this paper we study the prenegotiation process of eliciting the negotiators' preferences and building the negotiation offer scoring system. We analyze how the agents build the formal and quantitative scoring systems based on the preferential information provided by their principals. The results of the bilateral negotiation experiment conducted in Inspire negotiation system are analyzed, in which the simple direct rating technique (SMARTS-like approach) is implemented to evaluate the negotiation problem and build scoring systems. The concordance of such scoring systems with the principal's preferences was determined using the cardinal and ordinal inaccuracy measures. Then for each agent the scoring system was determined using UTASTAR method based on the same preference structures subjectively declared for direct rating. Finally, the inaccuracy of scoring systems obtained by means of both methods was compared.

Keywords: Prenegotiation preparation · Negotiation offer scoring systems · Preference analysis · Direct rating · Holistic preference elicitation · UTASTAR

1 Introduction

Multiple Criteria Decision Aiding (MCDA) techniques can be used in the negotiation analysis to support the negotiation parties in eliciting their preferences and building the formal negotiation offer scoring systems [12, 21]. Out of variety of decision support methods, there is a direct rating technique very often used to support negotiators in scoring system construction [8, 19]. This straightforward method that derives from SAW preference model [7] requires of the negotiators assigning the cardinal ratings to the various resolution levels (options) of negotiation issues that may be used to formulate the negotiation offers. Similarly to SMARTS method [1], the global score of any offer (alternative) is determined then as the sum of the ratings assigned to the

© Springer International Publishing AG 2017
M. Schoop and D.M. Kilgour (Eds.): GDN 2017, LNBIP 293, pp. 202–214, 2017.
DOI: 10.1007/978-3-319-63546-0_15

options that comprise this offer. This direct rating approach is regarded to be one of the easiest MCDA techniques [6], yet the recent researches report on various problems with efficient use of it to support negotiators in preference elicitation, related mostly to their cognitive capabilities and heuristic thinking [14, 20]. Since the decision makers vary in evaluating and effective use of different MCDA techniques depending on their personal characteristics [15], the question arises if implementing different supporting techniques for prenegotiation preparation could improve the decision making effects and the accuracy of representing the negotiators' preferences. An accurate scoring system allows the negotiators to understand the negotiation process as well as the scale of the concessions better, and makes them able to represent the real goals and the priorities of the principal. This requires implementing easier to use and more user-friendly decision support tools that would reduce the cognitive demand and will not result in a tiresome, unclear and complicated preference elicitation process.

The theory of MCDA offers a multitude of different methods and approaches to support the decision makers in eliciting their preferences [2]. Some of them, such as AHP, Evan Swaps or TOPSIS have been considered to be used in the negotiation context [9, 13, 21]. Yet, all of them, similarly to direct rating, require disaggregating the negotiation problem into its atomic elements, such as issues and options (i.e. the issues' resolution levels), end evaluating these elements separately, regardless of the offers they may comprise in the forthcoming negotiations. This seems to differ from the natural decision context the negotiators face during the negotiation process, in which they have to compare the subsequent offers submitted in a consecutive negotiation rounds in a form of full packages, i.e. defined as the compromise/contract proposals specifying the resolution levels of all negotiation issues. In our view, there are other MCDA approaches that offer the preference elicitation algorithms better fit to the specificity of the prenegotiation analysis. For instance, some holistic approaches aim at determining the quantitative scoring systems based on the general preferential information provided by the decision makers over the examples of offers (complete packages) in a form of the rank order or by sorting them into a predefined categories. There are examples of implementing holistic approaches to prenegotiation support, such as UTA method in the Mediator system [5], MARS [3] or calibrated ELECTRE-TRI [22]. Unfortunately, no study of use and usefulness of these approaches in negotiation context was conducted, nor the comparison of the accuracy of the results obtained by these methods and other MCDA techniques was made. Hence, no conclusions on the superiority (or inferiority) of the holistic approaches over the classic ones based on direct rating can be drawn.

The goal of our research is to find whether the holistic approach can be regarded as an efficient and effective method in negotiation support, i.e. if it allows building the scoring systems more (or at least not less) accurate than the scoring systems determined by means of direct rating technique (e.g. the SMART-like). We choose UTA (in particular its modified version of UTASTAR) [17] method as a representative of the holistic approach for the pragmatic and technical reasons. ELECTRE-TRI is a sorting method, and it does not allow creating the scoring systems of sufficient precision. On the other hand, MARS is computationally complicated. Quite the contrary, UTASTAR method seems technically simplest for implementation but at the same time fits the negotiation context best. It allows comparing full packages, which is what the negotiators have to do later on during the actual negotiation phase. It does not require the negotiator to operate with numbers or cardinal ratings directly, hence it reduces the cognitive demand when compared to classic SMARTS-like techniques. The negotiators need to order the pre-defined set of reference packages only, so they need solely to judge if one is better or equivalent to another. What is more, they may freely define such a set of reference packages, so they may use as the examples the offers they subjectively consider easy to compare.

To compare the accuracy of the scoring systems obtained by means of direct ratings and UTASTAR methods we analyze the dataset of bilateral negotiations conducted by means of Inspire negotiation system [8], in which the participants (agents) determined their scoring systems according to the principal's recommendations and using the direct rating technique. Based on the same preferential information provided by agents, we determine similar scoring systems utilizing the UTASTAR method, using various technical parameters such as the form of reference set or minimal scoring difference.

The paper is organized as follows. In Sect. 2 we briefly sketch out the direct rating and UTASTAR preference elicitation techniques. In Sect. 3 the experiment is described as well as the laboratory study is presented. We discuss also the issue of adequate representing the principals' preference systems by the agents and notions of scoring system accuracy there. In Sect. 4 we present the results in scoring system accuracy obtained for the Inspire experiment for both direct rating and USTASTAR techniques. We conclude in Sect. 5 with final remarks and future work.

2 Tools for Determining the Negotiation Offer Scoring Systems

To offer any support to the negotiators the negotiation problem needs to be structured first and then scored according to their individual preferences. Structuring the negotiation problem requires identification of the problem, defining the objectives/issues, formulating the alternatives and their consequences and may be conducted according to selected MCDA approaches, such as PrOACT [4]. Such formal definition and structure of the problem is called a negotiation template [12] and needs to be evaluated by the parties to produce their individual negotiation offer scoring systems used later for asymmetric and symmetric support [11, 23].

Let us consider the template described by m issues. For each issue $i = 1, \ldots, m$ the countable sets X_i of options (x_i^j) are defined $(j = 1, \ldots, |X_i|)$. The negotiation offer scoring system can be formally defined as $\left(\{w_i\}_{\forall i}, \{v(x_i^j)\}_{\forall i,j} \right)$, i.e. a system of cardinal ratings describing the issue importance (weights) w_i and the option values $v(x_i^j)$. If the issues are quantitative and continuous they could result in uncountable sets of option within the feasible ranges. Thus, it is recommended to identify for them the set of selected salient options only to make the process of template scoring easier. Based on the ratings determined for these salient options (and implicitly defined marginal value functions) any option from the feasible range can be scored, for instance by using the notion of interpolation among the ratings of salient options.

2.1   Direct Rating Approach

The simplest way to evaluate the template is to assign a cardinal rating to each of the template's element in a way that would describe the decision maker's strength of preferences for the issues and options respectively. Usually the scale used is interval, and the cardinal ratings describe the relative importance of the negotiation template elements. The direct rating procedure that allows building the negotiation offer scoring systems consists of the following two steps:

Step 1. The negotiator assigns ratings (weight) to each of the issues such as:

$$\sum_{i=1}^{m} w_i = 100. \tag{1}$$

Note, that there may be other pool of points used to build the scoring system than 100 used in formula (1). Yet, the 0-100 scale seems to be most popular.

Step 2. The negotiator evaluates each option $x_i^j \in X_i$ within each negotiation issue i by assigning the rating $v(x_i^j)$ such as:

$$v(x_i^j) \in \langle 0; w_i \rangle, \text{ for } i = 1, \ldots, m \text{ and } j = 1, \ldots, |X_i| \tag{2}$$

The most preferred (best) option obtains the maximum possible rating, i.e. w_i; while the worst – the rating equal to 0.

The global rating of any offer $a \in A$ that can be built based on the options defined within the template (i.e. identified in the sets X_i) is determined as an additive aggregate of ratings of options that comprise this offer:

$$V(a) = \sum_{i=1}^{m} \sum_{j=1}^{|X_i|} z_i^j(a) \cdot v(x_i^j), \tag{3}$$

where $z_i^j(a)$ is a binary multiplier indicating if the j th option of the i th issue was used to build the offer a (1) or not (0).

2.2   UTASTAR Holistic Approach

In the holistic approach a major role is played by the preference disaggregation principle, according to which the marginal value functions that describe the decision maker's preferences over the options of single criteria (issues) can be inferred out of their global preferences revealed for the complete alternatives. In UTASTAR, being an improved version of original UTA method [16], it is assumed that there is a subset A_R of reference alternatives (offers), for which the decision maker (negotiator) is able to define their preferences by building a ranking (partial order) of them. It is also assumed that the negotiator's preferences are monotonous, i.e. for the set of ordered options the value functions are monotonously non-increasing or non-decreasing. The latter condition may be easily waived by introducing some additional constraints into the model.

According to the UTASTAR algorithm, the rank order of alternatives from A_R declared by the negotiator is used to formulate the following linear program:

$$\min(z) = \sum_{k=1}^{|A_R|} [\sigma^+(a_k) - \sigma^-(a_k)] \tag{4}$$

subject to:

$$\Delta(a_k, a_{k+1}) \geq \delta, \text{ if } a_k \succ a_{k+1}$$
$$\Delta(a_k, a_{k+1}) = 0, \text{ if } a_k \sim a_{k+1}$$
$$\sum_{i=1}^{m} \sum_{j=1}^{\alpha_i - 1} w_{ij} = 1$$
$$w_{ij} \geq 0, \sigma^+(a_k) \geq 0, \sigma^-(a_k) \geq 0$$

where: $\sigma^+(a_k)/\sigma^-(a_k)$ – are the overestimation and underestimation errors for the rating of offer a_k, $\Delta(a_k, a_{k+1})$ is a difference in ratings for offers a_k and a_{k+1}, and $w_{ij} = v(x_i^{j+1}) - v(x_i^j)$ is a difference in ratings for two subsequent resolution levels of issue i.

By solving the linear program (4) the ratings $v(x_i^j)$ of each option of each issue are obtained. If alternative solution occur, some LP sub-problems are defined and solved to find the set of univocal ratings.

It is worth noting, that apart from the preference structure provided by the rank order of alternatives directly by the negotiator there are also other technical factors that may influence the final results of LP model. They are: the reference set A_R (the number and form of alternatives chosen to compare); δ coefficient describing the minimal differences between two alternatives in the rank order; and the parameter α_i, describing the number of salient options for each issue and simultaneously, the number of sections into which the marginal value function is divided into (precision of single-criteria evaluation). Hence, while analyzing the accuracy of scoring systems based on the UTASTAR approach a kind of sensitivity analysis should be conducted.

2.3   Scoring Systems Accuracy

The negotiation offer scoring systems are built to help the negotiators in evaluation of negotiation offers, scale of concessions and final compromises. To make these evaluations reliable, they should reflect the preferences of the negotiators in most adequate way. The MCDA techniques vary in cognitive demand and hence may cause different difficulties to the negotiators in adequate representing their intrinsic preferences. Consequently, the scoring systems built by means of different MCDA techniques may not be ideally concordant with the implicit preferences of the negotiators. The situation is more complicated if we consider the principal-agent context of the negotiations [18]. Is such a context, where there is an agent negotiating on behalf of their principal, there is another issue that requires special consideration. Namely, if the agents are purely focused on the principal goals and preferences and represent their preferences only in the scoring systems they build; or there are also some agents' goals taken into account that change the preference structures and make the scoring systems different from those determined based on pure principals' preferences. Thus, it is important to measure the concordance and accuracy between the scoring systems that can be determined by agents by means of different MCDA techniques and the reference scoring system of the principal.

Let us denote by S^P the scoring system reflecting the principal's preferences in the most precise way, and by S_t^A – the scoring system determined by the agent using technique t. The inaccuracy (discordance) of two scoring systems may be measured from the viewpoint of two different notions separately, as ordinal and cardinal inaccuracy. Based on the notions of Kendall rank correlation and Jaccard distance the ordinal inaccuracy may be measured as the number of pairs of elements of the negotiation template, for which the ratings in S_t^A preserve the same rank order as in S^P. Formally, the **ordinal inaccuracy index** is defined by the following formula

$$OI(S^P, S_t^A) = |L| - \sum_{l=1}^{|L|} r_l, \tag{5}$$

where L is a set of all pairs of the negotiation template elements that can be compared $(L = \frac{m(m-1)}{2} + \sum_{i=1}^{m} \frac{|X_i|(|X_i|-1)}{2})$; and r_l is a binary indicator describing concordance (1) or discordance (0) of the ranks resulting from ratings for l th pair in S^P and S_t^A. Note, that not all pairs of the template elements are compared while determining the OI index. The weights and the ratings of the options within each issue are compared separately.

The second inaccuracy measure, i.e. cardinal inaccuracy, will measure not the correctness of the rank orders between S^P and S_t^A, but the adequacy of reflecting the strength of preferences from S^P by S_t^A. Taking into account the fact that the issue weights are represented in the scoring systems by the ratings of most preferable options within each issue, determining the differences in ratings for both issue weights and maximum option ratings would result in double counting of agents' errors in defining their preferences. Hence, the **cardinal inaccuracy index** would take into account the differences in option ratings only:

$$CI(S^P, S_t^A) = \sum_{i=1}^{m} \sum_{j=1}^{|X_i|} |v^P(x_i^j) - v_t^A(x_i^j)|. \tag{6}$$

3 Experimental Setup

To compare the accuracy of UTASTAR-based negotiation offer scoring systems with the one determined by the negotiators using direct rating approach a research study was organized that consisted of two stages: experimental and simulation ones.

3.1 Online Negotiation Experiment – Stage 1

The experimental part of our study amounted to organizing the bilateral electronic negotiation experiment, which was conducted in Inspire negotiation system [8] in spring 2015. In the experiment 332 students from five countries (Austria, Canada, Netherlands, Poland and Taiwan) and six universities (IMC University of Applied Sciences, Krems; Carleton University, Ottawa; Radboud University, Nijmegen; University of Bialystok, University of Economics in Katowice and National Taiwan

Normal University) took part. In our experiment we used one of Inspire's standard business negotiation problems that concerned signing a contract between two agents representing: (1) the entertainment company (Mosico), and (2) the musician (Fado) [10, 14]. In this problem the negotiation template is defined by means of four issues, each with predefined set of salient options. A detailed private information is provided to each agent that specify the principal's preferences in both verbal and graphical form. The example of private info for Mosico party is shown in Fig. 1.

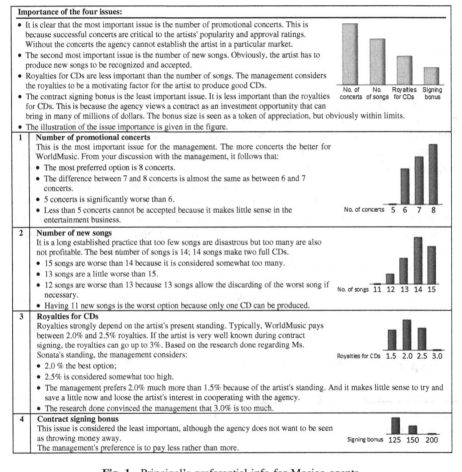

Fig. 1. Principal's preferential info for Mosico agents.

In the prenegotiation preparation phase organized in the Inspire system the participants were asked to determine their own subjective negotiation offer scoring systems using the direct rating approach (S_{DR}^A), as presented in Sect. 2.1. The negotiators declared the ratings in the input boxes corresponding to each element of the negotiation template. They entered the numbers directly, no sophisticated graphical support was offered at this stage. Neither the correctness of such scoring systems nor concordance with principals' preferences were checked by Inspire, so the agents were free in assigning the rating points according to their own interpretation of private info.

Having eliminated the incomplete records we have obtained a complete dataset of 126 negotiation instances (252 negotiators). To isolate various case-specific factors (such as the differences and nuances in principal's preference structures) that could influence the results of our experiment for further analysis we have selected the negotiators representing one party only, i.e. the Mosico agents.

3.2   Simulation Analyses – Stage 2

Second stage of our study aimed at: (1) determining for each negotiator an alternative scoring system by means of UTASTAR algorithm (S^A_{UTA}) and (2) comparing the accuracy of agents' scoring systems obtained by means of direct rating (S^A_{DR}) and UTASTAR algorithm (S^A_{UTA}) with the reference rating of the principal (S^P). The latter one was determined based on precise mapping the bars heights from principal's preferential info (Fig. 1) into the cardinal numbers. Since there is no UTASTAR-based rating mechanism implemented in the Inspire, the S^A_{UTA} systems could not be determined by the negotiators themselves during the online experiment. Therefore, we simulated their work using an Excel spreadsheet and an Excel add-in designed purposely for this analysis. For each negotiator we calculated their UTASTAR-based scoring system (S^A_{UTA}) using the same subjective preferential information provided in Inspire during the process of determining the scoring systems by means of direct rating (S^A_{DR}). Hence, we had implicitly assumed that if the negotiators had to use the UTASTAR preference elicitation algorithm themselves, they would have provided us with the preferential information perfectly concordant with the one provided for the purpose. This is a strong assumption, but if omitted it would accept the situation in which different amount of preferential information is used for building S^A_{DR}), and different for S^A_{UTA}. Yet, we need to emphasize that due to some cognitive limitations of the negotiators and some heuristics and biases that may occur during the preference elicitation process it is possible that the same decision maker (negotiator) would provide different preferential information if asked twice.

There were also a few technical issues to be considered before performing the analysis, i.e.: (1) which offers should comprise a set of reference alternatives A_R; (2) what should be the value of δ parameter; (3) how to handle the problem of non-monotonic preferences for issues of "Number of new songs" and "Royalties for CDs"?

We decided to use two alternative reference sets A_R. The first one was the same to the one displayed by Inspire system to the negotiators for the verification in the last stage of preference elicitation phase. This is a heuristically reduced set of orthogonal vectors, which consists of 13 alternatives (A^{13}_R) selected according to normal distribution rules (Table 1). The second reference set was comprised of all 25 orthogonal vectors that can be determined for Mosico-Fado case (A^{25}_R).

Table 1. Set of reference alternatives A_R^{13} as used in Inspire system.

No	Concerts	Songs	Royalties	Contract
1	5	11	1.5	125000
2	5	12	1.5	150000
3	5	15	1.5	125000
4	5	15	1.5	150000
5	6	12	1.5	150000
6	6	12	2.0	150000
7	6	13	2.5	200000
8	7	13	2.0	150000
9	7	13	2.0	200000
10	7	13	2.5	200000
11	7	14	3.0	125000
12	8	14	2.5	125000
13	8	14	3.0	125000

Considering the different possible δ values we decided to use the three following levels: 0.01; 0.02 and 0.05. Thus, we assume that the consecutive alternatives in the ranking provided by agents need to differ at least of 1, 2 or 5 rating points.

Finally, we solved the problem of non-monotonous preferences for the consecutive option values of two negotiation issues by using the ordinal scaling functions that changed the original options values into a new series that preserved the rank order described in the private info (Fig. 1). Namely, the set of options $\{11, 12, 13, 14, 15\}$ for the issue of number of new songs was mapped into a corresponding set: $\{5, 4, 3, 1, 2\}$; while for the options of royalties the following mapping was implemented: $\{1.5, 2.0, 2.5, 3.0\} \rightarrow \{3, 1, 2, 4\}$. For both sets of new option values the agent's marginal preference functions were assumed to be non-increasing. Hence, we implicitly assumed that the agents are ordinally accurate, i.e. the option ratings determined by UTASTAR algorithm preserve the rank order of options defined by the principal.

Using the above assumptions regarding the UTASTAR models for each agent the S_{UTA}^A scoring system was determined by solving an appropriate linear program according to model (4).

4 Results

The specificity of UTASTAR algorithm requires providing explicitly, what is the monotonicity of all marginal value functions used to determine the whole scoring system. As described in Sect. 3, we assumed that the marginal value function used in our analysis are concordant to those defined by the principal, and consequently, that our negotiators are ordinally accurate. Thus, our analysis of the differences in the scoring systems accuracy have to be limited only to those negotiators, whose $OI(S^P, S_{\text{DR}}^A)$ index is equal to 0 (those who really appeared to be ordinally correct in direct rating). This reduces the numbers of records from the Inspire dataset to 49 only.

We started the analysis of the scoring systems obtained by means of classic direct rating and the UTASTAR approach from comparing how different they can be for the same preferential information provided by the agents. We used formulas (5) and (6) to determine the $OI(S_{DR}^A, S_{UTA}^A)$ and $CI(S_{DR}^A, S_{UTA}^A)$ indexes (Table 2).

Table 2. Average ordinal and cardinal differences in S_{DR}^A and S_{UTA}^A of accurate Mosico agents for various parameters of UTASTAR model.

δ	A_R^{13}		A_R^{25}	
	$\overline{OI}(S_{DR}^A, S_{UTA}^A)$	$\overline{CI}(S_{DR}^A, S_{UTA}^A)$	$\overline{OI}(S_{DR}^A, S_{UTA}^A)$	$\overline{CI}(S_{DR}^A, S_{UTA}^A)$
0.01	2.1	87.1	0.02	10.6
0.02	2.0	75.1	0.02	8.1
0.05	1.9	57.3	2.10	56.9

We found that when the set A_R^{13} is used the rating systems S_{UTA}^A and S_{DR}^A are quite different. Various δ values did not change the average ordinal difference $(\overline{OI}(S_{DR}^A, S_{UTA}^A))$ significantly ($p = .000$ for all pairs in Wilcoxon test), yet from the viewpoint of average cardinal difference $\overline{CI}(S_{DR}^A, S_{UTA}^A)$ the higher δ values the more similar S_{UTA}^A and S_{DR}^A are (all average CI indexes differ significantly for $p > .157$). There are however quite different results obtained for S_{UTA}^A scoring system determined out of the full orthogonal set A_R^{25}. These systems seem to be significantly better than S_{UTA}^A determined out of A_R^{13} for $\delta = \{0.01; 0.02\}$ with respect to average ordinal and cardinal difference ($p = .000$ in Wilcoxon test), but $\delta = 0.05$ makes the granularity of differences too big and the similarity of S_{UTA}^A and S_{DR}^A scoring systems appears to be as poor as for A_R^{13}.

The final issue was the accuracy of S_{UTA}^A and S_{DR}^A when compared to S^P. We determined the average ordinal and cardinal inaccuracy indexes for the S_{DR}^A scoring systems of 49 Mosico agents $(\overline{OI}(S^P, S_{UTA}^A) = 0$ and $\overline{CI}(S^P, S_{UTA}^A) = 35.8$ respectively) and compared them with the corresponding indexes for the S_{UTA}^A scoring systems obtained for different UTASTAR models. The average ordinal and cardinal inaccuracy indexes for various S_{UTA}^A scoring systems are shown in Table 3.

The results show, that S_{UTA}^A obtained for A_R^{13} and different δ values are significantly less accurate than the S_{DR}^A scoring systems when the same preferential information (input data) is used. For the analysed 49 ordinally accurate Mosico agents (each with $OI(S^P, S_{DR}^A) = 0$) the S_{UTA}^A scoring systems appeared to have, on average, two rank orders changed across the whole negotiation template when compared to S^P $(\overline{OI}(S^P, S_{UTA}^A) \approx 2$). The average cardinal inaccuracies for S_{UTA}^A (94.3; 82.6 and 64.3) are also significantly higher than the corresponding cardinal inaccuracy of S_{DR}^A (35.8). The significance in differences is confirmed for $p = .000$.

The results appear a little different for the inaccuracies of S_{DR}^A and S_{UTA}^A obtained out of A_R^{25}. For two δ values equal to 0.01 and 0.02 the average ordinal accuracy of S_{DR}^A and S_{UTA}^A does not differ significantly ($p = .327$). Additionally, for $\delta = 0.02$ the average cardinal inaccuracy $\overline{CI}(S^P, S_{UTA}^A)$ differs only of 2.7 rating points from

Table 3. Average ordinal and cardinal inaccuracy of S_{UTA}^A with respect to S^P of Mosico agents for various parameters of UTASTAR model.

δ	A_R^{13}		A_R^{25}	
	$\overline{OI}(S^P, S_{\text{UTA}}^A)$	$\overline{CI}(S^P, S_{\text{UTA}}^A)$	$\overline{OI}(S^P, S_{\text{UTA}}^A)$	$\overline{CI}(S^P, S_{\text{UTA}}^A)$
0.01	2.1	94.3	0.02 *	40.8
0.02	2.0	82.6	0.02 *	38.5
0.05	1.9	64.3	2.10	70.2

inaccuracy $\overline{CI}(S^P, S_{\text{DR}}^A)$ – 38.5 vs. 35.8; yet, the Wilcoxon test confirms the difference to be significant at $p = .000$. This shows it is possible to achieve the inaccuracy level very close to the ones obtained in direct rating by manipulating with various parameters of UTASTAR model.

5   Conclusions

In this research study we aimed at finding if UTASTAR-based holistic approach may be efficacious in determining the negotiation offer scoring systems. We compared the accuracy of scoring systems obtained by means of different UTASTAR models with the ones determined by means of direct rating approach. UTASTAR seem to be less cognitively demanding, since it requires of the negotiators to define their preferences using the rank orders only instead of operating with cardinal scores. We were afraid, however, that it might have an impact on scoring systems accuracy.

As we found, the results depend strongly on the parameters used to setup the UTASTAR models. The form of the reference set A_R influences the average accuracy of S_{UTA}^A scoring systems and their similarity to the corresponding S_{DR}^A scoring systems. Generally, the full orthogonal set A_R^{25} results in more accurate scoring system that the holistically reduced one (A_R^{13}). Too high granularity in differences of the alternatives from A_R (δ value) may also have a negative impact on the accuracy of the S_{UTA}^A scoring systems, especially when the number of the alternatives in A_R is large and the accumulated differences may not fit the whole rating scale. This is a situation of A_R^{25} and $\delta = 0.05$, in which at least five alternatives need to be considered as equally good (assigned with the same rank) to satisfy all conditions in model (4).

Using A_R^{25} we were able to prove that the scoring systems obtained by means of UTASTAR technique may be no less accurate than the one determined by means of direct rating approach (for at least ordinal comparisons). Yet, the high number of alternatives the agent would need to compare themselves while building a required rank order for the UTASTAR model could be discouraging and tiresome, and result in errors related to heuristic-based thinking. In our laboratory study we also assumed, that the agent would be able to define their preferences for the purpose of UTASTAR algorithm as they did for direct rating. However, while analyzing the complete packages in holistic approach they may forget about the nuances in detailed structure of preferences at the level of options, and hence provide quite different preferential information.

Therefore our future work will be focused on identifying other reference sets, that would be less numerous but simultaneously consisting of the packages sufficiently diverse, which would allow UTASTAR algorithm to capture adequately the nuances of single criteria rating functions and produce more accurate scoring systems. We would also aimed at verifying our laboratory result in an experimental study, in which the agents would use UTASTAR-based negotiation support tools to generate their scoring systems and evaluate subjectively the use and usefulness of such a scoring approach.

Acknowledgements. This research was supported by the grant from Polish National Science Centre (2015/17/B/HS4/00941).

References

1. Edwards, W., Barron, F.H.: SMARTS and SMARTER: improved simple methods for multiattribute utility measurement. Organ. Behav. Hum. **60**(3), 306–325 (1994)
2. Figuera, J., Greco, S., Ehrgott, M. (eds.): Multiple Criteria Decision Analysis: State of the Art. Springer Verlag, Boston (2005). doi:10.1007/b100605
3. Górecka, D., Roszkowska, E., Wachowicz, T.: The MARS approach in the verbal and holistic evaluation of the negotiation template. Group Decis. Negot. **25**(6), 1097–1136 (2016)
4. Hammond, J.S., Keeney, R.L., Raiffa, H.: Even swaps: a rational method for making trade-offs. Harvard Bus. Rev. **76**(2), 137–149 (1998)
5. Jarke, M., Jelassi, M.T., Shakun, M.F.: MEDIATOR: towards a negotiation support system. Eur. J. Oper. Res. **31**(3), 314–334 (1987)
6. Keeney, R.L.: Decision analysis: an overview. Oper. Res. **30**(5), 803–838 (1982)
7. Keeney, R.L., Raiffa, H.: Decisions with Multiple Objectives: Preferences and Value Trade-Offs. Wiley, New York (1976)
8. Kersten, G.E., Noronha, S.J.: WWW-based negotiation support: design, implementation, and use. Decis. Support Sys. **25**(2), 135–154 (1999)
9. Mustajoki, J., Hamalainen, R.P.: Web-HIPRE: global decision support by value tree and AHP analysis. INFOR J. **38**(3), 208–220 (2000)
10. Paradis, N., Gettinger, J., Lai, H., Surboeek, M., Wachowicz, T.: E-negotiations via inspire 2.0: the system, users, management and projects. In: de Vreede, G.J. (ed.) Group Decision and Negotiations 2010 Porceedings, The Center for Collaboration Science, University of Nebraska at Omaha, pp. 155–159 (2010)
11. Raiffa, H.: The Art and Science of Negotiation. Harvard University Press, Cambridge (1982)
12. Raiffa, H., Richardson, J., Metcalfe, D.: Negotiation analysis: the science and art of collaborative decision making. The Balknap Press of Harvard University Press, Cambridge (2002)
13. Roszkowska, E., Wachowicz, T.: Application of Fuzzy TOPSIS to scoring the negotiation offers in ill-structured negotiation problems. Eur. J. Oper. Res. **242**(5), 920–932 (2015)
14. Roszkowska, E., Wachowicz, T.: Inaccuracy in defining preferences by the electronic negotiation system users. In: Kamiński, B., Kersten, Gregory E., Szapiro, T. (eds.) GDN 2015. LNBIP, vol. 218, pp. 131–143. Springer, Cham (2015). doi:10.1007/978-3-319-19515-5_11

15. Roszkowska, E., Wachowicz, T.: Analyzing the applicability of selected MCDA methods for determining the reliable scoring systems. In: Bajwa, D.S., Koeszegi, S., Vetschera, R. (eds.) Proceedings of the 16th International Conference on Group Decision And Negotiation Bellingham, Western Washington University, pp. 180–187 (2016)
16. Siskos, Y., Grigoroudis, E., Matsatsinis, N.F.: UTA methods. In: Multiple Criteria Decision Analysis: State of the Art Surveys, pp. 297–334. Springer, New York (2005). doi:10.1007/0-387-23081-5_8
17. Siskos, Y., Yannacopoulos, D.: UTASTAR: an ordinal regression method for building additive value functions. Investigaçao Operacional 5(1), 39–53 (1985)
18. Spremann, K.: Agent and principal. In: Bamberg, G., Spremann, K. (eds.) Agency Theory, Information, and Incentives, pp. 3–37. Springer, Heidelberg (1987)
19. Thiessen, E., Shakun, M.: First nation negotiations in Canada: action research using SmartSettle. In: Kilgour, D.M., Wang, Q. (eds.) Proceedings of Wilfried Laurier University (2009)
20. Vetschera, R.: Preference structures and negotiator behavior in electronic negotiations. Decis. Support Sys. 44(1), 135–146 (2007)
21. Wachowicz, T.: Decision support in software supported negotiations. J. Bus. Econ. 11(4), 576–597 (2010)
22. Wachowicz, T.: Negotiation template evaluation with calibrated ELECTRE-TRI method. In: de Vreede, G.J. (ed.) Group Decision and Negotiations 2010, The Center for Collaboration Science, University of Nebraska at Omaha, pp. 232–238 (2010)
23. Young, H.P.: Negotiation analysis. University of Michigan Press, Ann Arbor (1991)

The Heuristics and Biases
in Using the Negotiation Support Systems

Gregory Kersten[1], Ewa Roszkowska[2(✉)], and Tomasz Wachowicz[3]

[1] Concordia University, Montreal, Canada
gregory@jmsb.concordia.ca
[2] University of Bialystok, Faculty of Economy and Management,
Warszawska 63, 15-062 Białystok, Poland
erosz@o2.pl
[3] University of Economics in Katowice, Department of Operations Research,
1 Maja 50, 40-287 Katowice, Poland
tomasz.wachowicz@ue.katowice.pl

Abstract. In this paper we analyze the problem of recognizing the cognitive heuristics, in particular the errors of perception and information processing, and their impact on the activities of negotiators undertaken in the prenegotiation phase to define, structure and analyze the negotiation problem. We focus on evaluating and analyzing the impact of scaling biases on the accuracy and concordance of negotiation offer scoring systems with the preferential information provided to negotiating agents by their principals. In our study we use the dataset of bilateral electronic negotiations conducted by means of Inspire negotiation support system, which provides users with decision support tools for preference based on direct rating approach. The results of experiments confirm the necessity of building the heuristics-sensitive decision support tools for negotiation support.

Keywords: Preference elicitation · Negotiation offer scoring systems · Direct ratings assignment · Ratings accuracy · Biases · Heuristics

1 Introduction

Negotiation analysis is a research methodology [13] that provides a set of formal models, methods, algorithms and approaches for analyzing and supporting the negotiation, mediation and arbitration processes. They are very often used in the negotiation support systems (NSS) to facilitate the parties in making the rational decisions on rejecting or accepting the negotiation offers, analysing the negotiation process, measuring the scale of concessions and evaluating the efficiency of the negotiation compromise [6]. On the other hand, the results of researches in experimental economy emphasize the decision makers' (DM) limited rationality and common using of intuition and heuristics instead of conducting systemic and rational decision analysis while making various managerial decisions. As described by psychologists, the human brain works using two separate system of thinking. System 1, called "fast thinking", operates automatically and quickly, with little or no effort and no sense of voluntary control while

© Springer International Publishing AG 2017
M. Schoop and D.M. Kilgour (Eds.): GDN 2017, LNBIP 293, pp. 215–228, 2017.
DOI: 10.1007/978-3-319-63546-0_16

analyzing the facts, reasoning and finding the answers (solutions). It is mostly based on the intuition, connotations and heuristics. System 2, called "slow thinking", embodies the rational and analytical approach for problem solving, allocates attention to the effortful mental activities that demand it, including complex computations [3, 16]. Heuristics are simple cognitive procedures that allow to solve the problems quickly, though not always adequately and precisely enough, or answer the questions that appear [4, 18]. Thus, the heuristics should be taken into account while analyzing DMs' actions, moves and decisions in situations, in which not only the rationality but also the emotions, time pressure or other non-content-related factors play the role.

The experimental results show that the DMs and the negotiators quite often use the System 1 apply a wide range of heuristics to find the solution of the decision problem under consideration [4] or in planning their negotiation strategy and conducting the negotiation talks [9, 12]. This may results in very many groundless and unjustified assumptions and biases.

It seems scientifically challenging and vitally important to verify the virtual efficacy of the decision support methods applied in NSSs in eliminating the negative effects of heuristics-based thinking. The necessity of such a verification appears to be more evident in the view of the results of recent research conducted in a field of electronic negotiation. They prove that the NSS users are, in majority, inaccurate in defining their preferences and mapping the principal's preferential information into the formal and cardinal scoring system determined by means of direct rating methods [15].

In this paper we present the results of preliminary research of the ongoing project focused on recognizing and evaluating the potential impact of heuristic-based thinking on the prenegotiation activities. We focus analyzing the impact of scaling biases on the accuracy of the negotiation offer scoring systems and their concordance with the preferential information provided to negotiators to the negotiators by their principals or institutions on behalf which they negotiate. We use the dataset of bilateral electronic negotiations conducted by means of Inspire negotiation support system [7]. The paper consists of three more sections. In Sect. 2 we discuss the problem of heuristics and biases in the negotiation support systems. In Sect. 3 we describe the Inspire experiment, while in Sect. 4 we provide general results concerning scaling biases observed in the experiment. We conclude in Sect. 5 with some comments on the future work.

2   Heuristics, Biases and Negotiation Support

Electronic negotiation is conducted by means of software support tools that offer the decision support modules [1, 7]. The latter are implemented to help negotiators focusing on the negotiation problem more analytically. These modules stem from the theory of multiple criteria decision making (MCDM) and provide negotiators with formal algorithms and procedures they may be used in the prenegotiation phase to elicit their preferences and determine the negotiation offers scoring systems. Based on these scoring systems the negotiators may conduct a detailed and rational analysis of the potential negotiation alternatives (i.e. offers and agreement proposals), measure their profitability, value the concessions and analyze a fairness of the negotiated agreement. Such a support should initiate and stimulate the negotiators to uses System 2 of analytical thinking and eliminate this way a biased and heuristics-based decision making style.

Unfortunately, the NSS decision modules usually apply the simplest method of eliciting preferences based on direct rating assignment. This is a version of SMARTS algorithm [2] based on simple additive weighting (SAW) preference model [5] that derives directly from the theory of rational choice and multiple attribute value theory. However, some researches indicate the problems with correct assigning the rating points and misinterpretation of global ratings of offers the users have while operating with scoring systems based on direct rating approach [15, 21].

That the problem of cognitive limitations and heuristics in negotiation, such as the anchoring or framing effects, is well described in a literature on the theory of negotiation [12]. However, it was not analyzed in a context of analytical activities the negotiators need to perform in the prenegotiation phase, i.e. during the process of eliciting their preferences and determining the negotiation offer scoring systems. There are only few research works, in which the necessity of considering the negotiators' bounded rationality in the negotiation analytics or the design of the negotiation support systems is risen [17, 22]. The problem of modification and tuning of the existing MCDM methods, such as SMARTS, AHP [11], ELECTRE [19] or TOPSIS [14] to the specificity of the negotiation process and cognitive capabilities of negotiators was not studied neither.

The recent authorial researches [15, 20] indicate that the heuristic-based thinking accompanies the prenegotiation process supported by means of formal MCDM methods. They show that the way the preferential information are presented (*framing effect*), as well as the reference points chosen for the further preference analysis play the key role in the negotiation problem structuration. They also report numerous mistakes made during the process of building the negotiation offers scoring system that result from misperception of the verbal and graphical preferential information, misusage of the scoring points and misinterpretation of the scoring system obtained this way. Hence, the detailed studies on using the heuristics and intuition in negotiation analytics should be conducted, which would allow to answers the questions on the impact of heuristic-based errors and mistakes made in prenegotiation phase on the negotiators' further decisions and actions at the later stages of the negotiation process.

3 Experimental Setup

To verify the existence of heuristics in prenegotiation analytics the bilateral negotiation experiment organized in 2014 and conducted in Inspire negotiation support system was analyzed. The participants were 332 students from Austria, Canada, Netherlands, Poland and Taiwan. We studied the biases that had appeared during the process of building the negotiation offer scoring system by means of direct rating procedure implemented in Inspire. In the experiment the Mosico-Fado case was used, in which a contract between the agents of entertainment company – Mosico, and the singer – Fado is negotiated. The negotiation template was defined by four issues, each having a predefined list of salient options (see Table 1). The preferences of both Mosico and

Table 1. Mosico-Fado negotiation template.

Issues to negotiate	Options
Number of new songs (introduced and performed each year)	11; 12; 13; 14 or 15 songs
Royalties for CDs (in percent)	1.5; 2; 2.5 or 3%
Contract signing bonus (in dollars)	$125,000; $150,000; $200,000
Number of promotional concerts (per year)	5; 6; 7 or 8 concerts

Fado principals were clearly described verbally and graphically and provided to the agents as private info (for details see [15]).

In the experiment the agents used the direct rating approach to determine their individual (subjective) negotiation offer scoring systems on the basis of the principal's preference info. The direct rating procedure implemented in Inspire consists of two straightforward steps:

(1) defining the issue weights (*issue ratings*), i.e. assigning the weights to each issue I_j $(j = 1, \ldots, m)$ in a form of cardinal ratings so that

$$\sum_j u_j = 100. \tag{1}$$

(2) defining preferences for options within each issue (*option ratings*), i.e. assigning the ratings u_{jk} to each option $x_{jk} \in X_j$ within each negotiation issue j so that

$$u_{jk} \in \langle 0; u_j \rangle, \tag{2}$$

and the most preferred (best) option receives the maximum possible score (i.e. u_j), while the worst – the rating equal to 0.

The global rating $u(A)$ of any offer A can be determined as the sum of ratings assigned to each option that comprise this offer, i.e.:

$$u(A) = \sum_{j=1}^{m} \sum_{k=1}^{|X_j|} z_{jk}(A) \cdot u_{jk}, \tag{3}$$

where $z_{jk}(A)$ is a binary multiplier denoting if the option x_{jk} comprises an offer A (1) or not (0).

The concordance of negotiators' individual offer scoring systems with the principals' preferences defined in private info may be measured by means of two notions of ordinal and cardinal accuracy (for details see [15]). In this study we will utilized the concept of ordinal accuracy only, which is focused on analyzing the concordance of the rank orders in the agent's scoring system and principal's reference scoring system. The agent's scoring system consists of the following set of weights and option ratings

$$S_A = \left(\{u_j\}_{\forall j}, \{u_{1k}\}_{\forall k}, \ldots, \{u_{mk}\}_{\forall k} \right), \tag{4}$$

defined for all elements of the negotiation template. Thus, in each scoring system there is $m+1$ sets of various ratings, one describing issue weights, and m describing the option ratings within each issue. A corresponding principal's reference scoring system S_P can be built separately for each negotiation party that precisely reflects the principal's preferences defined in private info (see Table 2). The notion of ordinal accuracy of agent's scoring system would require all the sets of ratings to reflect the same rank orders to the ones defined in the principal's scoring system. For instance, if the issue weights in the principal's scoring system are ordered according to non-increasing preferences, i.e.: $u_1^P \geq u_2^P \geq \ldots \geq u_m^P$ the ordinal accuracy requires the agent's issue ratings (weights) u_j^A to satisfy the following condition

$$u_1^A \geq u_2^A \geq \ldots \geq u_m^A. \tag{5}$$

Table 2. Principal's radius-based reference scoring systems for Mosico and Fado party.

Party	Reference principal's ratings															
	No. of concerts				No. of songs					Royalties for CDs				Contract bonus		
	5	6	7	8	11	12	13	14	15	1.5	2.0	2.5	3.0	125	150	200
Mosico	0	21	26	32	0	7	16	28	21	13	23	16	0	17	10	0
Fado	32	25	21	0	0	8	20	32	24	0	7	12	16	0	15	20

Similar conditions need to be satisfy for all sets of option ratings. Otherwise we say that scoring system is ordinary inaccurate.

Since the principal's preferences were visualized in private info graphically by means of circles, the principal's reference scoring systems may be determined by measuring either the radiuses or areas of circles. As we refer mainly to ordinal relationship between ratings, we will use one of these reference systems, namely the radius-based one (see Table 2). Note, that the issue weights u_j are defined here implicitly and can be derived according to formula (2) as $u_j = \max_k u_{jk}$ for each $j = 1, \ldots, 4$.

4   Results

4.1   Biases and Related Errors

While analyzing the experimental results we may observe various errors made by the negotiators in scoring the negotiation template, resulting in bigger or smaller discrepancies between their own scoring systems and the reference scoring system of the principal. Not all of them may result from the cognitive biases, but can also have a motivational background. Please note, that we analyze the results of negotiations conducted in principal-agent's context; hence, if the agents are driven by their own, subjective goals they would purposely determine different scoring systems than the ones that fit the principal's priorities only. We believe that the errors we observed in our experiment are rather related to cognitive biases, since in our experiment the participants were students, and the final grades they obtained for this assignment depended (among others) on how good contract they negotiated for their principal.

In our analyses we focused on the scaling biases, i.e. the ones which occur when the scales used by negotiators (agents) for ratings options or issues are mismatched. Within the scaling biases the following examples of detailed biases can be distinguished [10]:

- **omission bias** - some important issues/options were overlooked by the negotiator and they focus on some salient issue/options only,
- **contraction bias** - underestimating large differences and overestimating small differences between the importance of negotiation issues/options,
- **equalizing bias** - tendencies to allocate similar ratings to all issues/options.

The consequence of the above biases is an incomplete problem descriptions resulting from the agent's oversimplified mental model. The omission bias results in narrowing the negotiation space (generating a subset of feasible offers only) and hence some promising tradeoffs or creative compromises are simply excluded from the consideration. Similar effects may be observed for contraction bias, for which only good option/issues are exposed and other marginalized in building alternative negotiation offers. In general, the scaling biases may lead to misperception of offers' value and poor understanding of the negotiation process.

To recognize the occurrence of the scaling biases in the experiment we analyzed the agents' scoring systems using different statistical measures, such as minimum, maximum, and average. The results are presented in Tables 3 and 4.

Table 3. The statistics for option ratings for Fado and Mosico agents.

Statistic	Option ratings in Inspire experiment															
	No. of concerts				No. of songs				Royalties for CDs				Contract bonus			
	5	6	7	8	11	12	13	14	15	1.5	2.0	2.5	3.0	125	150	200
Mosico																
Min	0	0	0	0	0	0	0	0	0	0	0	0	0	0	0	0
Max	30	50	80	85	23	35	40	50	40	40	30	30	40	59	67	40
Av.	0.2	17.9	29.9	36.9	0.34	9.9	18.6	28.2	17.8	7.7	15.5	10.8	2.5	9.5	8.1	1.9
Fado																
Min	0	0	0	0	0	0	0	1	0	0	0	0	0	0	0	0
Max	42	38	40	34	30	45	58	70	65	28	30	35	40	40	81	74
Av.	31.3	24.7	17.2	0.6	0.5	11.8	22.1	32.8	23.8	0.2	7.6	12.0	13.7	1.13	14.1	17.2

The analysis of the basic statistics confirm the occurrence of errors that may result from scaling biases, i.e. the omission and contradiction biases. We can observe that with respect to all options and across all issues the minimum ratings are equal to 0 and

Table 4. The statistics for issue ratings (weights) for Fado and Mosico agents.

Weights	Fado				Mosico			
	Concerts	Songs	Royalties	Contract	Concerts	Songs	Royalties	Contract
Min	5	1	1	5	8	12	1	1
Max	42	70	40	81	85	50	40	67
Av.	32.67	33.63	14.75	19.00	39.90	29.9	17.48	12.57

the maximum ratings overestimate the references ratings even more than three times. For instance, minimum issue weight for Royalties is equal to 1, and the maximum – 40, while the reference weights are equal to 23 and 16 for Mosico and Fado respectively.

To track the occurrence of the biases in the Inspire experiment we define the following errors that address some of the scaling problems described in [10]:

- **Error 1.** The agent's rating of one issue is at most 5, while other is rated at least 50; or the issue weight is equal to 1 (marginalized).
- **Error 2.** The not-worst option from reference system is rated as 0 by the agent. This Error may be broken down into three others:
  - **Error 2a.** The worst option from reference system is not rated as 0.
  - **Error 2b.** At least two options are rated as 0.
  - **Error 2c.** The worst option from reference system is not rated as 0 and at least two other options are rated as 0.

It is easy to prove the following relationships among errors:

- The worst option from reference system is rated as 0 (Error 2a) implies that the not-worst option from reference system must be rated as 0 (Error 2).
- At least two options are rated as 0 (Error 2b) implies that the not-worst option from reference system is rated as 0 (Error 2).

Please note, that the errors that we defined above do not indicate the occurrence of different scaling biases disjunctively. E.g. the Error 2 may address the occurrence of the omission bias as well as the contraction and equalizing biases. Yet, our goal (as defined in introduction) is to study the occurrence of scaling biases in general, and hence we will not investigate the detailed relations among the errors and biases defined.

Let us also emphasize that the definition of Error 1 is subjective (the rating levels are defined arbitrarily), and here the way of preference visualization can influence on ratings assigned individually by agents to the issues. This would require further investigation and comparison with other visualization techniques. Yet, the definition of Error 2 seems universal in a sense that it should be independent of way of visualization because it is based on the simple rule addressing the order of preferences only: the recognition of the not-worst option and the worst option in the reference system.

4.2   Errors Observed in Inspire Negotiation Experiment

As shown in Figs. 1 and 2, Error 1 was observed for 5,1% of Fado agents and 13% of Mosicos.

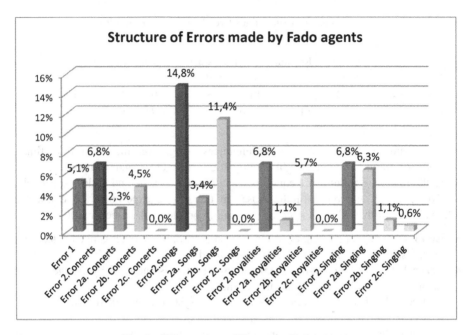

Fig. 1. The structure of Errors for Fado agents.

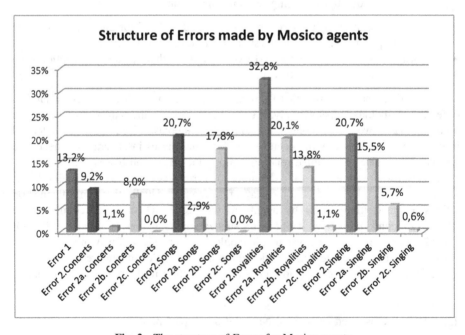

Fig. 2. The structure of Errors for Mosico agents.

The Fado agents overrated mainly the issues of Songs and Contract and underrated the Royalities, while Mosico agents overrated Concerts and underrated Contract. It is worth noting, that the Error 2 occurred most frequently for Songs (14,8%) for Fado agents, and in case of Mosico agents – for Royalities (32,8%), Contract (20,7%) and Songs (20,7%). At least two zeros (Error 2b) have been assigned most frequently for the options of issue of Songs (11.4% of Fado agents and 17.8% of Mosicos). The worst option from reference system was not rated as 0 (Error 2a) most frequently for options of Contract (6.3% of Fado agents and 15.5% of Mosico agents). What is also interesting, that the representatives of Mosico agents seemed to be more bias-prone that the representatives of Fado. It may be related either to the specificity of the structures of preferences provided by Mosico and Fado principals or to the intrinsic characteristics of the participants that played each role (e.g. demographic and decision-making profile) and requires further investigation in future research.

Deriving form Figs. 1 and 2 we may learn of the occurrence of specific errors for Fado and Mosico agents while analyzing the subsequent elements of the negotiation template. Yet, it is interesting to learn how often the heuristics occurred for each experiment participants. The structure of occurrence of errors for both negotiation roles are given in Fig. 3.

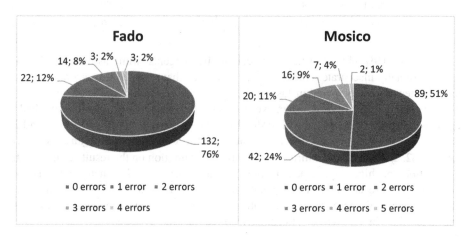

Fig. 3. The structure of Errors' occurrence for Fado and Mosico agents.

There is as much as three quarters of all Fado agents that did not reveal any error during the prenegotiation preparation phase, while nearly 50% of Mosico agents made at least one error resulting from the heuristic-based thinking. These differences are puzzling and suggest once again the necessity of conducting an in-depth analysis of behavioral profiles of the participants.

We have also analyzed what are the links between the errors revealed by the negotiators and the ordinal accuracy of the scoring systems they build. Generally, the problem of defining the ordinally accurate scoring systems seems to be quite common [15]. In our experiment, there were only 31 (18%) out of 176 Mosico agents who built their own scoring systems in perfect ordinal concordance with the principal's reference system (see Sect. 3, formula (5)). The percentage of ordinally accurate scoring systems for Fado agents is higher and equal to 22% (38 out of 174). The fraction test confirms the difference in ordinal accuracy of Fado and Mosico agents to be insignificant ($z = 0.99, p = 0.34$). Let us observe that Error 2 entails ordinal inaccuracy (see Eq. 4), but not conversely. Having analyzed the structure of relationships between the ordinal accuracy and defined errors the following results were obtained (Tables 5 and 6).

Table 5. Number of Fado agents with regard to scoring system accuracy and errors made.

	Ordinally inaccurate	Ordinally accurate	Sum
Errors 1 or 2	42	0	42
No Errors	94	38	132
Sum	136	38	174

Table 6. Number of Mosico agents with regard to scoring system accuracy and errors made.

	Ordinally inaccurate	Ordinally accurate	Sum
Errors 1 or 2	85	2	87
No Errors	60	29	89
Sum	145	31	176

There is 30.9% (42) of ordinally inaccurate Fado agents, while as much as 58.6% (85) of ordinally inaccurate representatives of Mosico that revealed Errors 1 and/or 2 in their prenegotiation preparation tasks. The difference is notable and statistically significant. What is interesting, when we compare the structures of Mosico and Fado agents who made or not made with ordinally accurate scoring systems (0 vs. 38 and 2 vs. 29) the fraction test for small samples would confirm both structure as equal ($\chi^2 = 0.62, df = 1$). This confirms our intuitive presumption on the results of thorough preparation and high engagement in prenegotiation activities. No matter what is the problem structure and nuances of principal's goals and preferences, the agents who have an analytical approach to prenegotiation appear not made heuristic-based errors and determine (in vast majority) ordinally correct scoring systems.

4.3   Other Errors Observed

Finally, we defined another error that could be directly linked to the ordinal inaccuracy of agents and explain the differences in fractions for roles (94 Fado agents vs. 60 Mosico agents, see shaded boxes in Tables 5 and 6). We checked the structures of accuracy in representation of the principal's preferences, as reported in our earlier work [8]. It appears that some elements of the negotiation template make bigger problems in assigning concordant ratings for agents, than others. In our experiment, for three elements of the negotiation template the non-monotonous preferences were defined by principals in private info (see Table 2), i.e.: for options of "No. of concerts" (both principals), and for options of "Royalties" (Mosico). For these three elements the percentage of agents with correct ratings is lower (29%–56%) than for all remaining elements of template (68%–79%), but one (Fado issue weights). It may suggest that the accuracy in agent's ratings could depend on the structure of preferences defined by the principal, and is higher when the principal's preferences are monotonous and lower for non-monotonous ones.

The last element of the negotiation template that made problems in accurate rating was the issue importance defined by Fado principal. Here, an order of last two issues described in private info (i.e. "Contract signing bonus" and "Royalties") was different from their order used in direct rating procedure, as shown in Fig. 4.

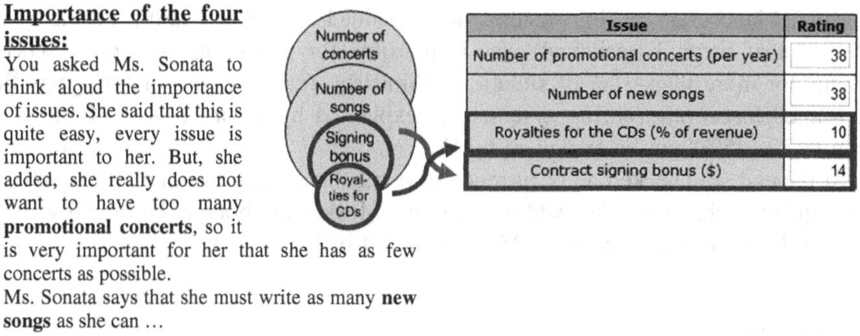

Importance of the four issues:
You asked Ms. Sonata to think aloud the importance of issues. She said that this is quite easy, every issue is important to her. But, she added, she really does not want to have too many **promotional concerts**, so it is very important for her that she has as few concerts as possible.
Ms. Sonata says that she must write as many **new songs** as she can …

Issue	Rating
Number of promotional concerts (per year)	38
Number of new songs	38
Royalties for the CDs (% of revenue)	10
Contract signing bonus ($)	14

Number of concerts
Number of songs
Signing bonus
Royalties for CDs

Preferencial info   **Direct rating procedure**

Fig. 4. Differences in listing issues in prenegotiation phase in Inspire. Source: [8]

It appeared, that only 60 out of 174 (35%) of Fado were able to assign the ratings that correctly.

Taking into account these additional problems we defined two complex errors separately for Fado and Mosico that take into account the nuances in principal's preference definition and prenegotiation organization of data by the decision support system (they may be related to such heuristics as an unintentional blindness):

- **Error 3 (Fado):** $u_{\text{Royalties}} \geq u_{\text{Contract}}$ or $u_{\text{Songs}} \neq u_{\text{Concerts}}$ or $u_{\text{Songs},13} \geq u_{\text{Songs},15}$,
- **Error 3 (Mosico):** $u_{\text{Royalties},1.5} \geq u_{\text{Royalties},2.5}$ or $u_{\text{Songs},13} \geq u_{\text{Songs},15}$.

We compared the frequency of occurrence of Error 3 with more general Errors 1 and 2. We found that for 125 Mosico agents (71%) the Error 3 occurs. 68 of them simultaneously made errors related to Error 1 or 2. For Fado party, 124 of agents (71%) made errors related to Error 3, but only 38 of them had did them together with errors related to Error 1 or 2. The relationships between the Errors are shown in Table 7.

Table 7. The relationships between errors for innacurate agents.

No. of respondents	Fado	Mosico
	Error 3	Error 3
Errors 1 or 2	38	68
No Errors 1 or 2	86	57
Sum	*124*	*125*

It is worth noting that the Error 3, representing the potential errors related to technical and role-related issues, was quite efficient in explaining the reasons for which the agents generated inaccurate scoring systems. For 86 Fado agents Error 3 occurred but not Errors 1 and 2. When we compare this number with 94 Fado agents, which determined inaccurate scoring systems not explained by Errors 1 and 2 (Table 5) we will find that Error 3 (technical) was responsible for generating inaccurate scoring systems for more than 91% of situations. Similarly, for Fado, out of 60 agents that determined inaccurate scoring systems not explained by Errors 1 or 2 (Table 6) as much as 57 (95%) can be explained by Error 3.

At the end we checked an occurrence of particular scaling bias, i.e. the equalizing bias. Equal weights were observed only for one Mosico and nobody Fado. All weights between 20-30 we observed for 5 Mosico and 14 of Fado agents.

5 Conclusions

In this paper we tried to verify the occurrence of heuristic-based thinking in the negotiators' activities in prenegotiation, mainly during the process of preference elicitation and determining the negotiation offer scoring system. Unexpectedly, despite being offered with the decision support, which aims at enforcing DMs to act analytically and rationally, the negotiators appeared to make different mistakes characteristic to fast and heuristic thinking. Some of the biases seem to be related directly to the intrinsic characteristics of agents related to their cognitive capabilities, number sense etc. that affect their understanding of principal's preferences (Errors 1 and 2), while other are related to some technical difficulties and support nuances unnoticed due to the heuristic-based fast thinking (Error 3). Yet, they indicate the necessity of organizing the decision support in negotiation in more thorough way that would be adjusted to the cognitive and perceptional capabilities and behavioral profiles of the negotiators. It seems that the selection of an adequate formal support package (method, algorithm, procedure, protocol) for facilitating the decision and analytical process in prenegotiation should be preceded by the detailed analysis of the demographical, negotiation and decision making profile of the NSS user and take into account their skills and educational background. As the results show, the assumption that the users are ultra-rational, and hence not prone to make mistakes in, one would think, such a simple task as assigning the ratings according to specific rules described by the principal, is simply incorrect. Thus, we should accept the fact that even if the negotiators (DMs) are engaged in the activities enforcing them to use the system of slow thinking, they would still act heuristically. The new negotiation support tools should take this phenomenon into account to be able to support them efficiently.

Acknowledgements. This research was supported by the grant from Polish National Science Centre (2016/21/B/HS4/01583).

References

1. Brzostowski, J., Wachowicz, T.: NegoManage: A System for Supporting Bilateral Negotiations. Group Decis. Negot. **23**(3), 463–496 (2013)
2. Edwards, W., Barron, F.H.: SMARTS and SMARTER Improved simple methods for multiattribute utility measurement. Organ. Behav. Hum. Dec. **60**(3), 306–325 (1994)
3. Evans, J.S.B.: The heuristic-analytic theory of reasoning: extension and evaluation. Psychon. Bull. Rev. **13**(3), 378–395 (2006)
4. Gilovich, T., Griffin, D., Kahneman, D.: Heuristics and Biases: The Psychology of Intuitive Judgment. Cambridge University Press, New York (2002)
5. Keeney, R.L., Raiffa, H.: Decisions with Multiple Objectives: Preferences and Value Trade-Offs. Wiley, New York (1976)
6. Kersten, G.E., Lai, H.: Negotiation support and e-negotiation systems: an overview. Group Decis. Negot. **16**(6), 553–586 (2007)
7. Kersten, G.E., Noronha, S.J.: WWW-based negotiation support: design, implementation, and use. Decis. Support Sys. **25**(2), 135–154 (1999)
8. Kersten, G.E., Roszkowska, E., Wachowicz, T.: An Impact of Negotiation Profiles on the Accuracy of Negotiation Offer Scoring System - Experimental Study Multiple Criteria Decision Making, vol. 11 (2016, in press)
9. Korobkin, R., Guthrie, C.: Heuristics and biases at the bargaining table. Marq. L. Rev. **87**, 795 (2003)
10. Montibeller, G., Winterfeldt, D.: Cognitive and motivational biases in decision and risk analysis. Risk Anal. **35**(7), 1230–1251 (2015)
11. Mustajoki, J., Hamalainen, R.P.: Web-HIPRE: global decision support by value tree and AHP analysis. INFOR J. **38**(3), 208–220 (2000)
12. Neale, M.A., Bazerman, M.H.: Negotiating rationally: the power and impact of the negotiator's frame. Executive **6**(3), 42–51 (1992)
13. Raiffa, H., Richardson, J., Metcalfe, D.: Negotiation Analysis: The Science and Art of Collaborative Decision Making. The Balknap Press of Harvard University Press, Cambridge (2002)
14. Roszkowska, E., Wachowicz, T.: Application of fuzzy TOPSIS to scoring the negotiation offers in ill-structured negotiation problems. Eur. J. Oper. Res. **242**(5), 920–932 (2015)
15. Roszkowska, E., Wachowicz, T.: Inaccuracy in Defining Preferences by the Electronic Negotiation System Users. In: Kamiński, B., Kersten, G.E., Szapiro, T. (eds.) GDN 2015. LNBIP, vol. 218, pp. 131–143. Springer, Cham (2015). doi:10.1007/978-3-319-19515-5_11
16. Stanovich, K.E., West, R.F.: Individual differences in rational thought. J. Exp. Psychol. **127** (2), 161–188 (1998)
17. Sycara, K.P.: Multiagent systems. AI Mag. **19**(2), 79 (1998)
18. Tversky, A., Kahneman, D.: Judgment under uncertainty: Heuristics and biases. Science **185** (4157), 1124–1131 (1974)

19. Wachowicz, T.: Decision support in software supported negotiations. J. Bus. Econ. **11**(4), 576–597 (2010)
20. Wachowicz, T., Kersten, G.E., Roszkowska, E.: The impact of preference visualization and negotiators' profiles on scoring system accuracy. In: 27th European Conference on Operational Research EURO 2015, University of Strathclyde, Glasgow (2015)
21. Wachowicz, T., Wu, S.: Negotiators' strategies and their concessions. In: de Vreede, G. J. (ed.) Proceedings of The Conference on Group Decision and Negotiation 2010. The Center for Collaboration Science, University of Nebraska at Omaha, pp. 254–259 (2010)
22. Zeleznikow, J., Bellucci, E.: Building negotiation decision support systems by integrating game theory and heuristics. In: Proceedings of the IFIP International Conference on Decision Support Systems (2004)

Author Index

Printed in the United States
By Bookmasters